NIMBUS
- Everything Electrical

Knud Jørgensen and Sune Nielsen

English translation:
Charles Duffill

Kim Scholer, 1990

* Something is funny in this electrical system.

NIMBUS
- Everything Electrical

First published 2020 as *Nimbus – Alt om el* by Books on Demand GmbH, Copenhagen, Denmark.

This edition is an English translation by Charles Duffill, Perth, Australia.

Text and illustrations: Sune Nielsen and Knud Jørgensen.
Photos and drawings, marked F&N, are from Fisker & Nielsen Ltd. Denmark.
Cover and layout: Knud Jørgensen.

Publisher: BoD – Hellerup, Denmark
Printing: BoD – Norderstedt, Germany

ISBN: 978-87-4303-185-7

Contents

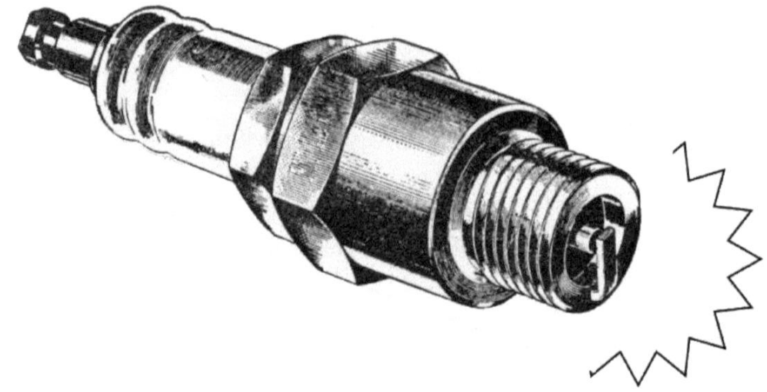

Introduction

Some motorcyclists are a little reluctant when it comes to dealing with electrical faults or with routine maintenance of the electrical system of their machines.

That's a pity, because it is really not so difficult, and it's absolutely essential for satisfactory motorcycling experience. So, in four sections, we will describe the Nimbus-C electrical system and outline what you - as the owner or rider - can easily do yourself when something goes wrong, and when it's best for you to seek advice or let the skilled folk take over.

It is important to keep in mind that all components of the electrical system depend on each other. No single part is unimportant, as everything is connected - literally, by cable.

Section 1: Wiring and switchgear

1.1 Wiring harness

The wiring harness of the Nimbus C consists of insulated multi-strand copper wires which have the task of conducting electrical current. The exact nature of that current is less important than a focus on practical issues - how it works for us, and why it sometimes doesn't.

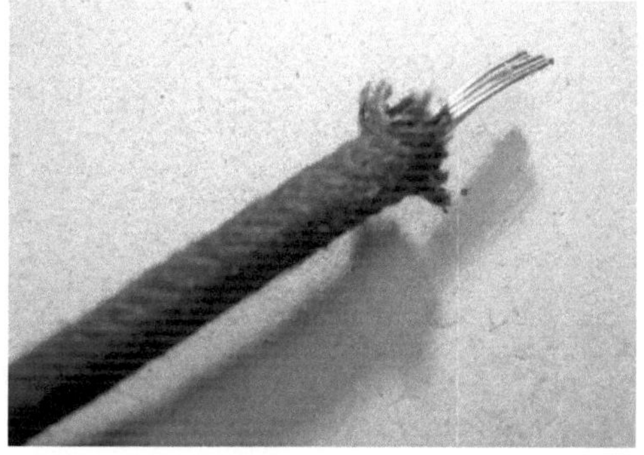

All electrical leads in a motorcycle wiring harness must have sufficient conductive capacity. This capacity depends on the nature and dimensions of the conductive material, and the required capacity depends on the operating voltage of the electrical system - 6 volts or 12 volts - and the electrical demands of the various components.

The Nimbus-C manufacturer (Fisker & Nielsen, Ltd.) specified that multi-strand copper cable with a cross-sectional area of 1.5 sq.mm should generally be used, except for the connection to the tail lamp, for which 1.0 sq.mm cable can be used. Cable of a lesser gauge than these must never be used, nor should cable intended for other than automotive purposes, such as house wiring, telephone or antenna connections.

The main earth lead, from the battery negative terminal to the motorcycle frame, is an important one. Experience suggests that it is good practice to use heavier gauge 2.0 or 2.5sqmm cable for this lead.

Automotive electrical cable in some markets is identified not by copper cross-section area (in sq.mm), but by nominal amperage. Cables suitable for Nimbus-C use are:
10Amp (1.1sqmm); 15**Amp** (1.8sqmm); 25**Amp** (2.9sqmm).

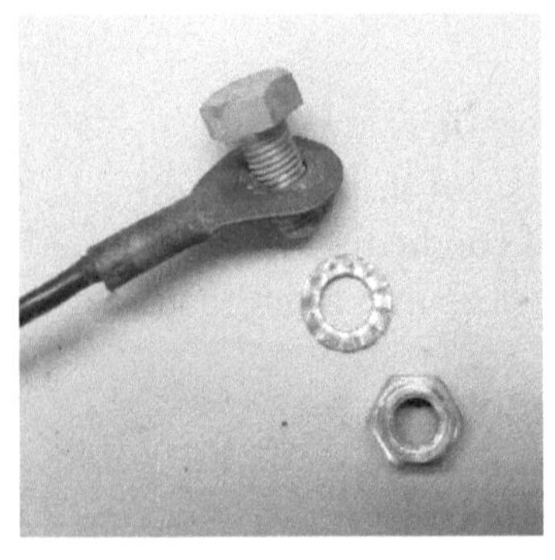

The battery earth connection to the frame needs to be secure and electrically sound. If the motorcycle has been repainted, then paint may need to be removed in order for good metal to metal contact. A star washer should be used between bolt head or nut and the frame.

Fibreglass mudguards and other fibreglass or plastic parts are non-conductive so cannot be used for earthing purposes.

Terminals are those parts used to attach cables to the battery, dynamo, regulator, and switches and so forth. In its simplest form, a terminal can be just a length of cable, stripped of insulation and formed into a loop secured by a twist or solder.

Other forms of terminal are either the factory brass originals, or available equivalents which may be intended to be soldered on, or crimped on with special purpose pliers.

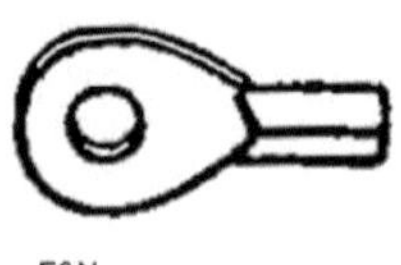

F&N

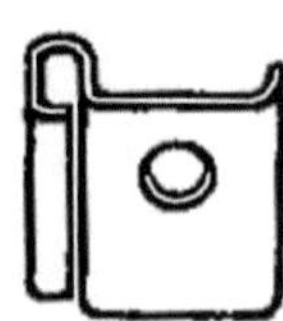

Wiring harness maintenance

It is a good idea to check all connections regularly, to ensure that terminals are in fact firmly attached to cables, and that there is no corrosion on the terminal or its attachment point.

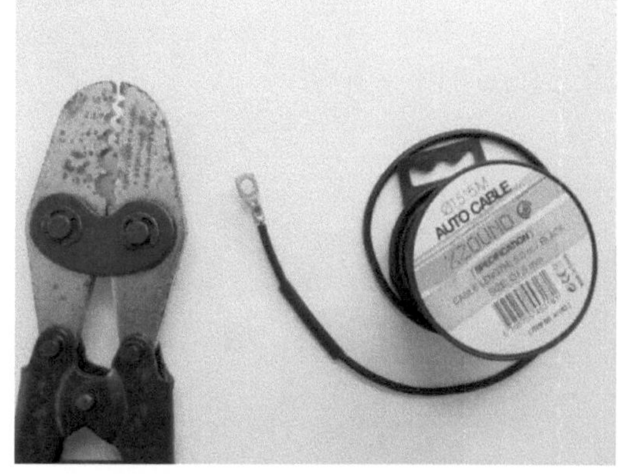

Terminals that are crimped on can work well if they are of the correct size for the cable and are attached using the correct crimping tool. But if ordinary pliers or a sidecutter has been used, a weak point may have been created in the system. Soldering can give a very secure connection, but sometimes this can be a false security as a major short circuit can melt solder in a few seconds.

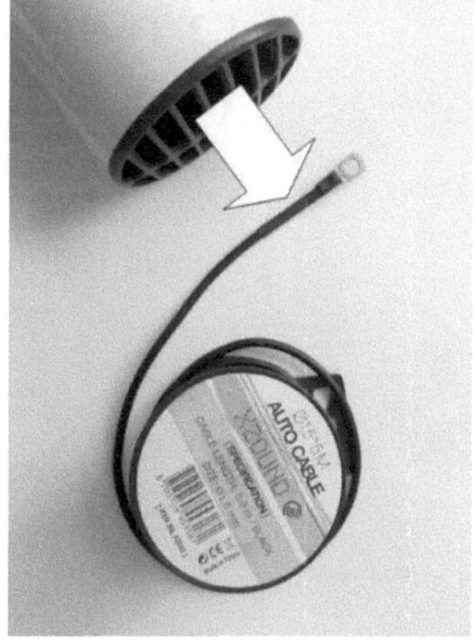

Shrinkfit tubing over the cable-terminal joint reinforces the connection, gives a good appearance, can be used for colour coding, and protects against moisture and corrosion.

The insulation of the electrical cables needs inspection and maintenance. The original insulation from the 1950's or earlier was fabric covered rubber and may have deteriorated too much to be kept in use. But more up to date cables can give problems too, as they must be of the correct specification, must be able to withstand mechanical strain, and must retain flexibility over time.

Regularly and systematically check the wiring harness. Be specially attentive to places where a cable might be strained or trapped or pulled tight over an edge.

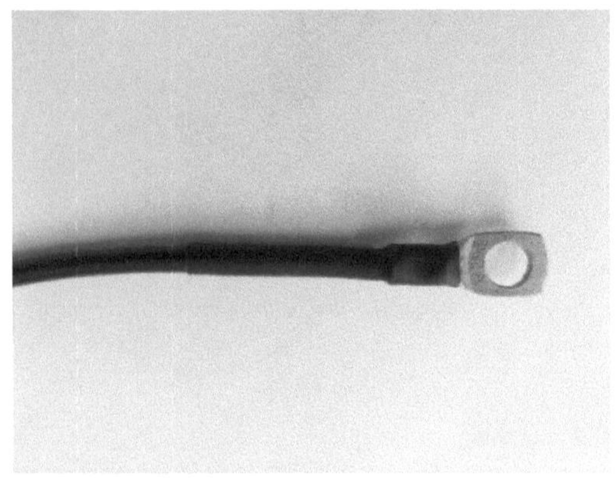

If there is any damage to cable insulation attend to this, as there is a risk of a short-circuit. Insulation tape can be a temporary repair. Shrinkfit tubing is more secure and long lasting, but the best solution is to completely replace the affected lead and to install it in a way which avoids future damage.

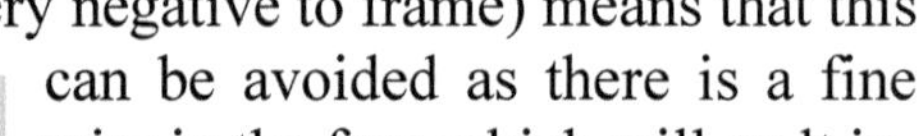

1.2 The Fuse

A short-circuit can in the worst case melt parts of the wiring harness in an instant. A fuse in the main earth lead (battery negative to frame) means that this can be avoided as there is a fine wire in the fuse which will melt instead.

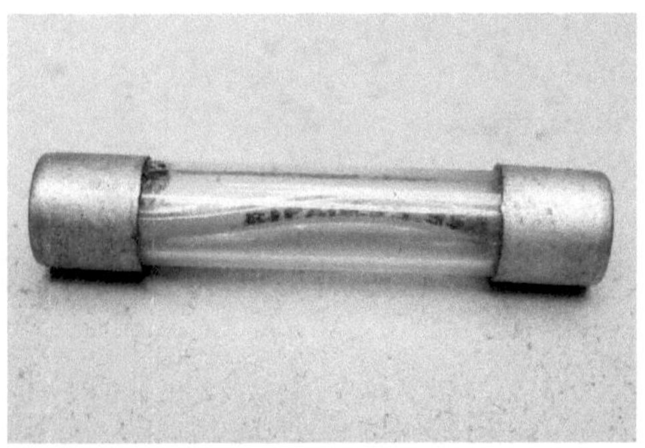

Until 1947 there was a 20-Ampere fuse holder provided in the main Nimbus earth lead.

For 1948 and later models it is advisable for the owner to add a fuse to the system if it has not already been done. This involves disconnecting the main earth lead from the battery negative terminal to the frame and inserting an in-line fuseholder in it.

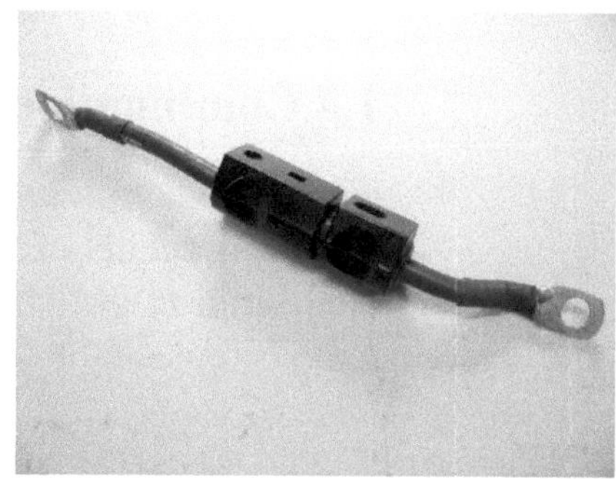

Fuse maintenance

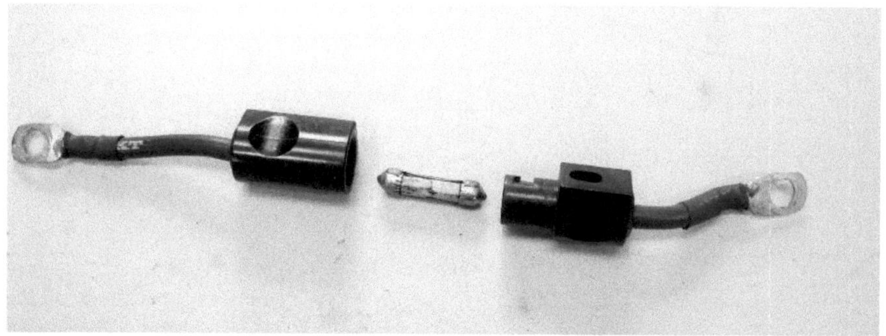

Regardless of how well the fuseholder parts fit, they may not be watertight. Moisture inside the fuseholder can result in corrosion on both the fuse and the internal fuseholder terminals. The fuseholder should be opened regularly to check this and to make sure that both terminals are securely attached to the earth lead.

1.3 Switches

Fortunately, there are few switches on a Nimbus, and they are well designed, durable, and repairable. Electrical problems may sometimes be due to nothing more than an easily rectified malfunction in the main handlebar switch - the "combination switch".

1.4 Combination switch

The combination switch at the handlebar is one of the brilliant technical details of the Nimbus-C. It combines the key-operated ignition switch and the twistgrip-operated lights switch and is connected to the horn button and the charge light.

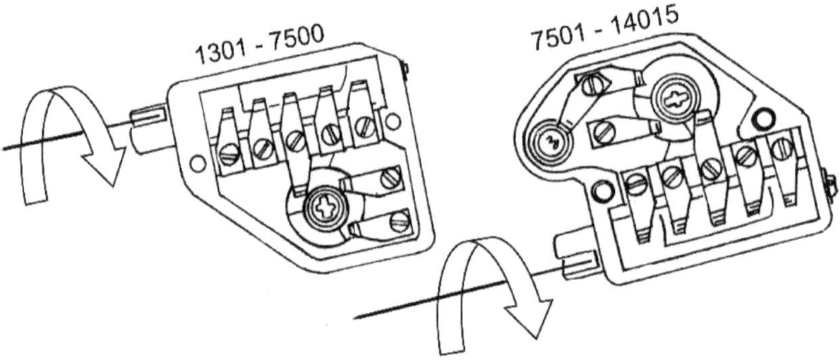

There are two versions of the combination switch block. The later one has the charge light built in. Both versions are attached to the underside of the pressed steel handlebar by two screws and can be removed from the left twistgrip with all wiring attached.

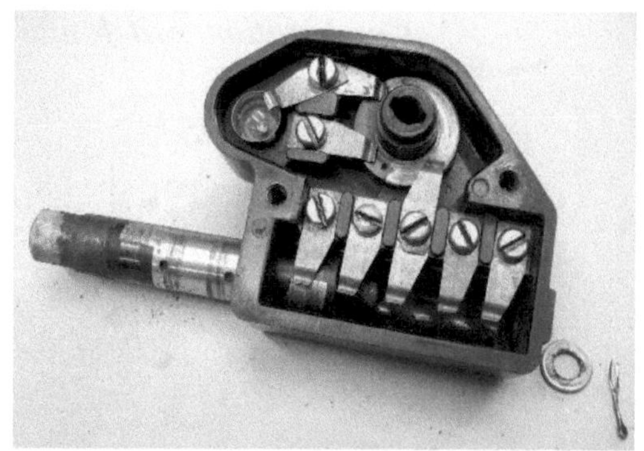

Both the ignition switch and the lights switch are rotary, with a number of brass leaf-spring wiper contacts.

The rotary ignition switch powers the ignition circuit and the lights switch. The rotor is made of bakelite (some replacement parts are nylon) with inlaid brass forming contacts. Over time, a faint metallic trace or track can develop between the contact areas of the rotor, with the possibility of a partial short circuit developing.

The lights switch rotor is made of ebonite which is non-conductive and not a particularly hard material. The switching action is a function of brass sections set into the rotor surface which make contact with different leaf-spring wipers as the rotor is turned by the left twistgrip. As in the ignition switch, a metallic track across the non-conductive surface may develop, and a partial short circuit may result.

Operation

When the key is simply inserted in the switch, nothing happens. But when the key - and therefore the rotor - is turned about a quarter turn clockwise, a connection is made from the battery positive to the lights switch and the horn button.
In this position, the key can be taken out.

When the key is turned a further quarter turn a connection is made from the battery positive to the ignition coil and to the charge light, which then turns on because it is connected to earth via the "D" terminal of the regulator.
In this position the key cannot be taken out.

Combination switch maintenance

A photo of the removed switch block with both rotors and all spring contacts in place will help correct reassembly.

Remove all leaf-spring contacts and take out the key rotor and main rotor. The main rotor is retained by a washer and split pin.

Using very fine emery paper carefully remove any metallic trace from the ebonite, cleaning the brass surfaces at the same time. (If the equipment is available, hold the rotor in a lathe or drill press chuck.) Clean the key rotor also.

Check every leaf-spring contact. If the contact area of any spring is excessively worn, replace the spring. Springs being reused can be given a little extra bend with long-nose pliers, to ensure that good contact is made with the rotor when all is assembled.

Charge light

The charge light is either fitted in the centre of the frame number disk in the handlebar (early models) or is incorporated in the combination switch block (later models).

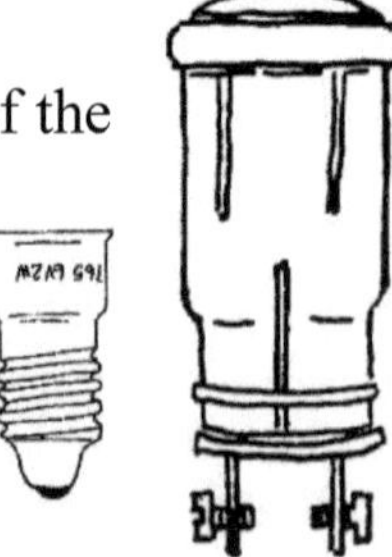

The charge light fitted in the handlebar usually requires no maintenance other than cleaning the terminals to ensure good electrical connection. The built-in charge light needs regular attention. When the key rotor is taken out, it is easy to remove the bulb. It is not screwed in, just held in place by one spring contact which unlike all others, has a corner cut off. Take out the bulb and clean off any deposits or corrosion. Also, remove the screw securing the lead to the D terminal on the underside of the switch block, and clean it thoroughly.

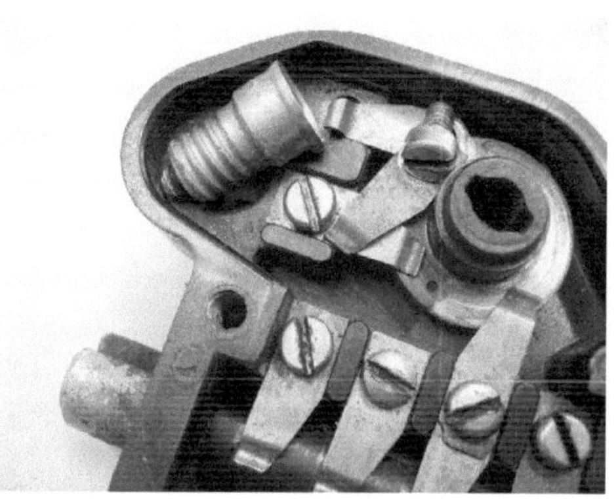

If the charge light bulb is heavily blackened this may indicate high system voltage. The bulb is likely to fail, and the problem with over voltage will need to be addressed. See Section 3.3 (Voltage regulator).

Horn button

The horn button is either a separate device mounted on top of the handlebar or it is built into the handlebar itself on the left side. From the horn button there is a lead to the horn. This lead can easily be trapped or pinched, which can cause a partial short circuit or voltage loss, with sound from the horn.

If the horn does not work correctly with a firm and definite press on the button, the problem is usually not with the switch, but with the connection to the horn or with the horn itself. See Section 2.5 (Horn).

1.6 Stoplight switch

This is located where it is exposed to moisture and road dirt, which can result in poor electrical contacts leading to voltage loss or short circuit. This could mean failure of the tail light or stoplight - a dangerous outcome.

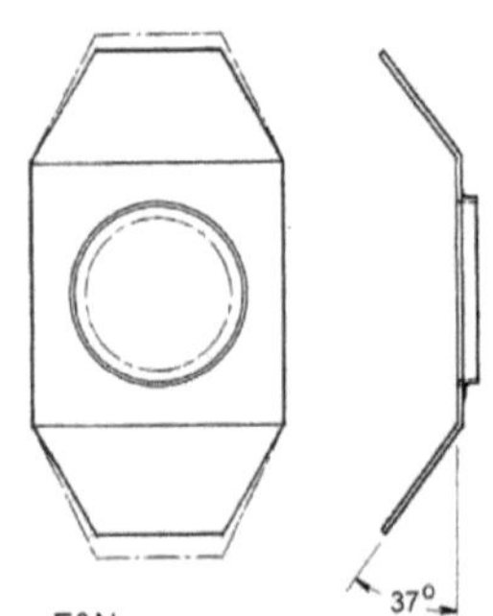

Stoplight switch maintenance

First, disconnect the battery earth lead. At the brake pedal, detach the tension spring which connects to the brake light switch to the brake pedal. Remove all cables, first labelling them B, S, and L. Remove the switch - the retaining nut is inside the frame.

Take the switch apart after noting how it looks when assembled. If possible, take a photo.

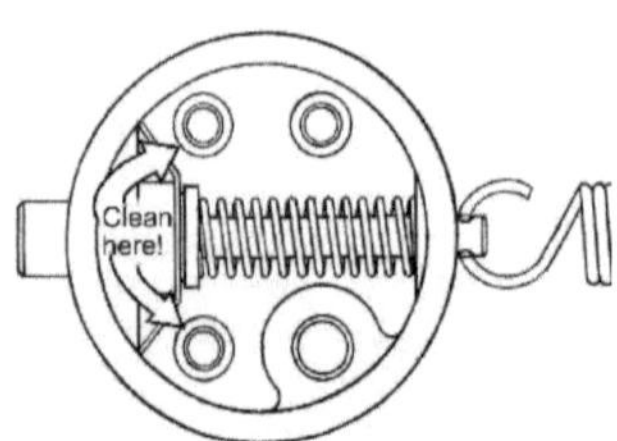

The contact leaf-spring, made of sheet brass, should be cleaned of any deposits or corrosion, and the same applies to the two threaded bushes which serve as switch contacts when the rear brake is applied. Sometimes it is also necessary to adjust the two bends of the spring.

Further, if the spring has been heated due to a short-circuit it can lose its temper - its spring property. If so, it must be replaced.

The internal compression spring returns the stoplight switch to off when the brake pedal is released and it may have become weakened. The free length of the spring when new was 35mm. If the free length now is less than this the spring may need replacing.

Section 2: Light and horn

"To see and to be seen"... This is surely a most important traffic safety rule.

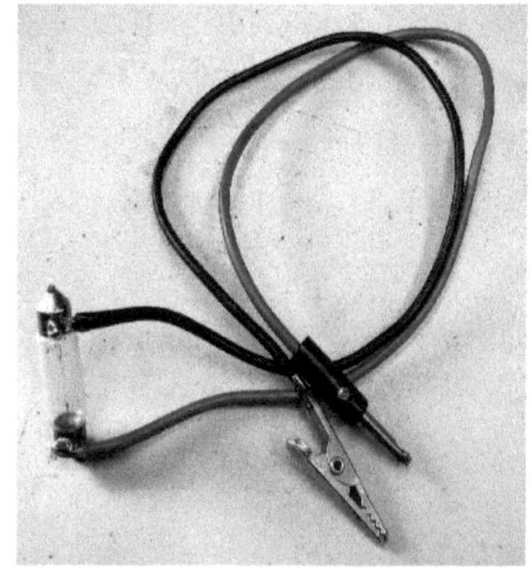

When inspecting or maintaining the lighting system a test light is convenient. One can easily be made from a bulb holder and spare lengths of cable, or one can be bought cheaply at a automotive accessory shop.

Ideally the test light should have a "crocodile clip" on one lead, for easy connection to the battery negative or a frame earth point, and a probe on the other.

2.1 Headlamp

The headlamp shell contains *the headlight* and *the parking light*. The glass lens of the headlamp must be complete and clean and, on the inside, dry. It must have a reflector which is neither tarnished nor has lost its reflective coating.

This has nothing to do with the electrical function, but everything to do with the light output.

2.2 Headlight

The headlight bulb has two filaments, one for low beam, the other for high beam.

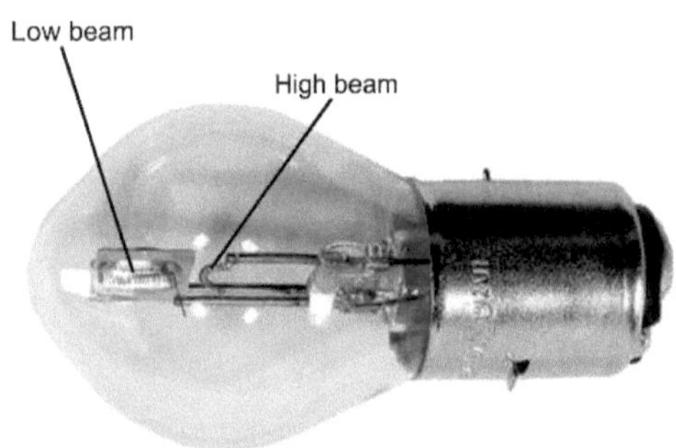

For both 6volt and 12volt systems, a 35/35-watt bulb is suitable. A bulb rated as 45/45-watt will be brighter but will draw a little more from the electrical system than is good for it. A 25-25-watt bulb is just bright enough for riding in daylight but is not noticeable by others in darkness.

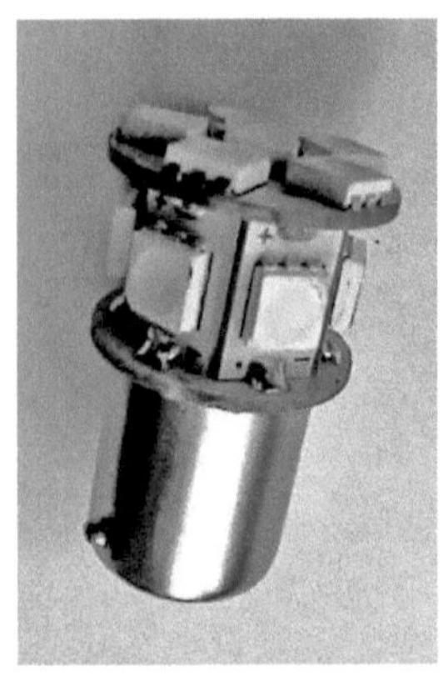

Other types of bulb or light sources have been developed which can be used in the Nimbus headlamp.

Halogen bulbs provide a brighter light for the same wattage or current usage. LED bulb replacements are available, offering superior light output with much reduced power consumption.

Note: There is a risk that an Insurance provider may consider use of LED lights to be a change of specification voiding insurance cover.

2.3 Parking light

The parking light is intended for use with an emergency stop at night and is fitted with a bulb of only 4W.
However, the 9.5mm socket will take a 20W halogen bulb. This is sufficiently bright to serve as a daytime running light (instead of using the headlight), reducing electrical demand by 15W. When fitting the bulb remember to hold it using an alcohol wipe or similar.

> *Note: An Insurance provider may consider use of LED lights to be a change of specification voiding insurance cover.*

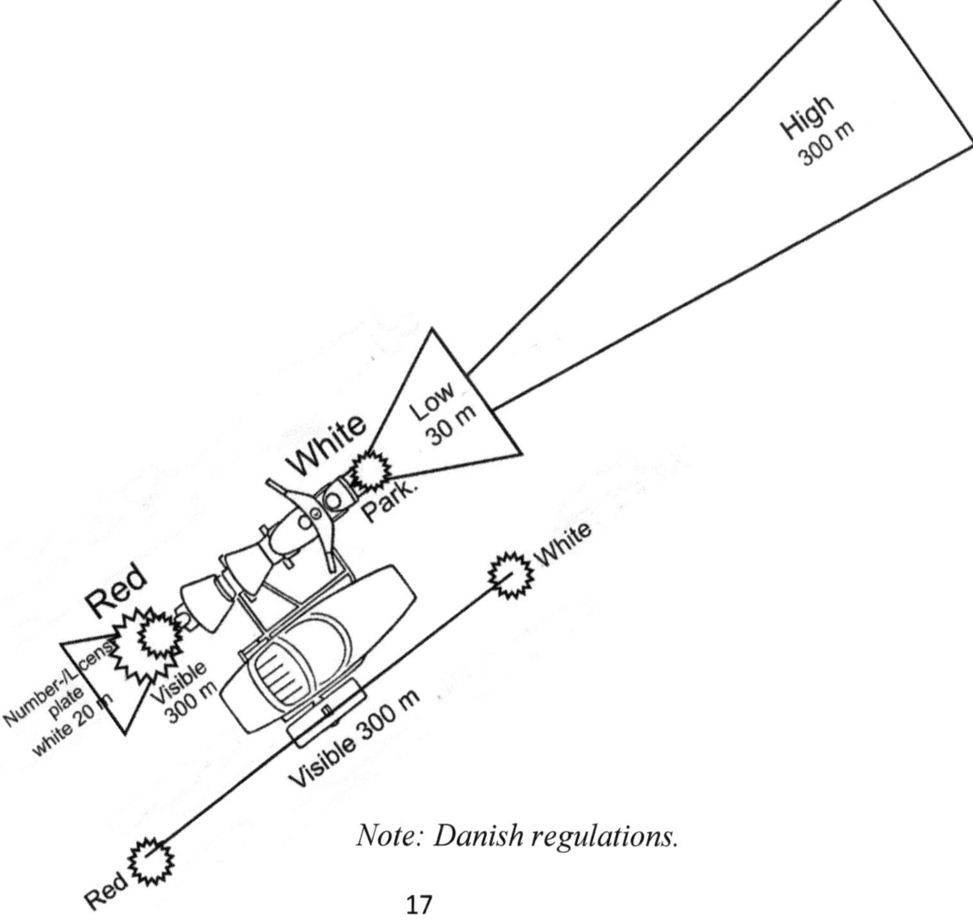

Note: Danish regulations.

Headlight maintenance

If the headlight bulb has darkened or a filament has failed, the bulb must be changed. If this happens repeatedly, check whether it is due to excessively high voltage in the system - see Section 3.3 (Voltage regulator) or to a short circuit within the headlamp. Also, the regulator earth connection may be relevant here. Check that this connection is clean and tight.

The headlight bulb socket has an earth lead, but it is connected only to the headlamp shell. With a newly painted headlamp or fork assembly, this may not be an effective earth connection as the fork assembly, being mounted in the steering head with 48 greased steel balls, may have poor metal-to-metal contact with the motorcycle frame.

Therefore it is advisable to add an earth lead direct from the headlight bulb socket, through the hole in the headlamp shell out to the frame earth point which is connected (through the fuse) to the battery negative terminal.

This extra earth lead is included in the ready-to-fit wiring harness available from Nimbus spares dealers.

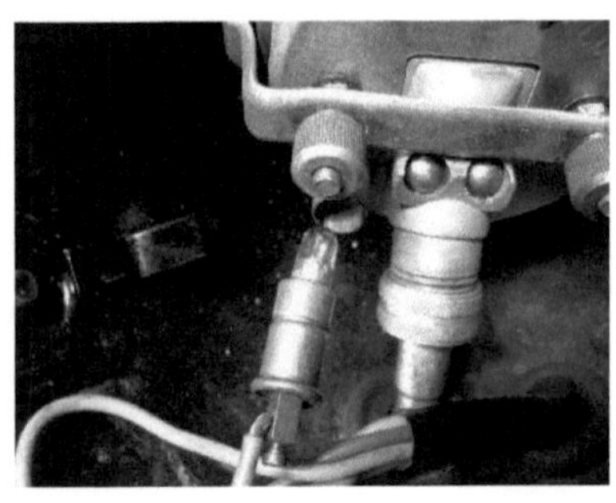

2.4 Speedometer light

If the speedometer is built into the headlamp, so is the speedometer bulb holder, as it fits directly into a socket in the body of the instrument.

Before Nimbus-C frame number 2551 the speedometer bulb holder was held in a clamp attached to the speedometer mounting bracket, between the speedometer and the ammeter.

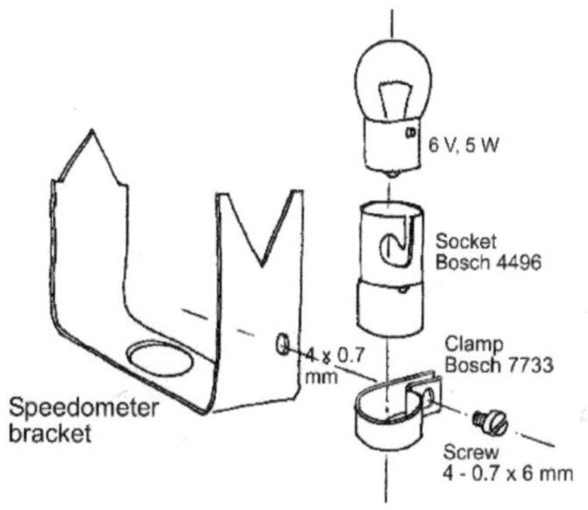

If the light is not working this is either because the bulb has failed, there is not a good connection in the handlebar combination switch, or there is no earth connection. Use a test lamp to locate the fault.

2.5 Rear lamp

The aluminium tail lamp casting is of the Nimbus factory's own manufacture. Some owners will have fitted an after-market unit with a larger or brighter light but the following advice relates to the original factory unit.

The rear lamp includes *the tail light* and a separate *stop light*, so there are two bulbs fitted behind the red reflective lens, which is marked JRU 129 or JRU 131.

There is an earlier version - in which the "lens" is a celluloid disc with red and yellow fields.

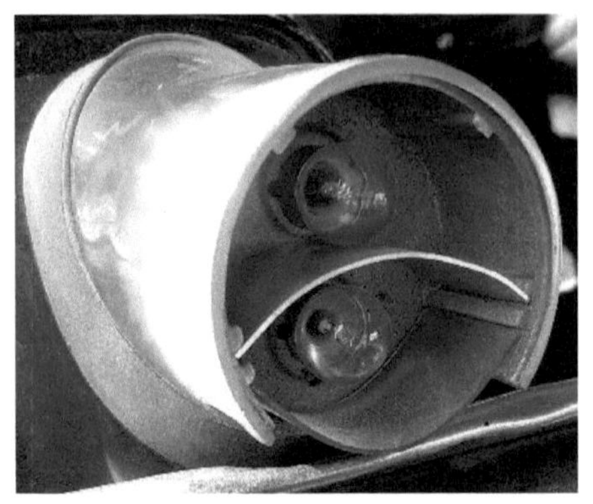

This does not meet current legal standards, and should not be used. The same applies to the old after-market lamp with the word "STOP" displayed. But! If using either of these non-reflective rear lamps, a round red reflector with a diameter of 65mm must be suitably fitted.

2.6 Tail light and stoplight

The tail light bulb is the lower one. It is a 5W bulb, showing to the rear through the red reflective lens, and at the same time illuminating the number plate through the clear panel below.

The stoplight bulb is the upper one. It is 15W, much brighter than the tail light.
Light from the two bulbs is separated by a curved aluminium insert.

Rear lamp maintenance
Both the tail light bulb and the stoplight bulb fit in sockets cast into the aluminium tail lamp housing. Only the pressure of the leaf-spring electrical contacts ensures good earth connections. This arrangement demands close attention.

Remove the large circlip which retains the lens and remove the red reflex lens. Press, hold down, and twist each bulb back and forth a little. If there is any sign of corrosion or deposits take out the bulbs and thoroughly clean both the bulbs and the leaf-spring contacts.

Use a test lamp to check each connection. If a bulb is loose in its socket it is a good idea to wedge it in place with a small piece of thin metal - cut, for example, from a drink can. Avoid wedging it so tight that it is difficult to remove.

The earth connection to the tail light depends on its mounting to the rear mudguard. Any new paint or corrosion there may well result in no such connection.

As in the case of the headlight, it is advisable to install a lead direct from the earth contact in the tail light to a location on the frame with a good earth connection, or to the battery negative terminal.

There is no form of reflector in the tail light and the dull grey of the aluminum casting is not very reflective, but there are remedies for this. Perhaps the easiest is to arrange some aluminium foil -silver paper - behind the bulb. Slightly crumpled, it works quite well. Another possibility is to paint the inside of the rear lamp white.

An LED stop-tail light insert, specially designed for the Nimbus 6V system, solves both the problem of brightness and longevity.

This insert is at present (2021) not on the market.

The advice given with respect to the headlight and parking light also applies here:

Note: An Insurance provider may consider use of LED lights to be a change of specification voiding insurance cover.

2.7 Sidecar light

The sidecar light bulb is so placed that it can be seen both from the front, as a white light, and from the rear, as red. There are types other than the "Ermax", which is shown here.

The sidecar bulb was originally required to be 5W, so that it could be seen from 300 metres, but if a 10W bulb is fitted, this is much better. Both the clear, forward-facing lens and the red rear lens must be 65mm in diameter. The red lens should be reflective, as in the Ermax, and marked JRU 130.

If the red lens is not reflective then a separate 65mm red reflector is required to be mounted nearby on the sidecar mudguard.

Sidecar light maintenance

The sidecar light lead is connected to the *tail light "L" terminal* of the stoplight switch. This long cable needs to be carefully secured to the sidecar frame tubes and thence to the underside of the mudguard. Because of likely exposure to road dirt and water, regular inspection and cleaning of the connection, the bulb, and the bulb holder is essential. Use a test light to check the circuit.

2.8 Horn

This is an important part of "a motor vehicle's signalling and warning equipment" and is required to be in working order.

The horn has a built-in electromagnet, activated when the horn button is held down, which attracts an iron armature fixed to a thin metal diaphragm. As soon as the armature is attracted to the electromagnet, power is automatically cut off allowing the diaphragm to spring back and close the electrical contact again, causing the process to be continually repeated until the horn button is released. The rapid oscillation of the diaphragm is perceived as a tone.

Over time, the Nimbus C has been fitted with horns from different suppliers, all with the same operating principle.

However some makers did not include a condenser, which prevents arcing at the electrical contacts in the horn.

Arcing, together with dirt and moisture, can result in a defective or non-working horn.

Horn maintenance

If the horn fails, begin by checking if there is power to it. Turn the ignition key to the first position. This turns on the 6-volt lead to the horn (from Terminal 5 on the combination switch). Use a test lamp connected to a good earth connection on the frame, or to the battery negative, and touch one of the terminals on the horn, and then touch the other.

If the test light comes on *in both_instances,* then the horn is probably not at fault and the problem may be the horn button.

If the test light comes on when connected to one terminal, *but not when connected the other,* then the contacts inside the horn might be the problem. Remove the front cover and clean the contacts thoroughly.

If the test light does not come on at all, check whether it lights when touching screw #5 on the underside of the handlebar combination switch. If it *does* light, then it is the lead from here to the horn which is faulty. If it *does not* light, then it is in the combination switch where the problem lies.

If the horn works, but the sound is too weak, some adjustment is possible by means of the slotted screw on the back of the horn. Take care to make only small changes at a time, so as not to move out of the adjustment range.

Section 3: Six volt system

3.1 Battery

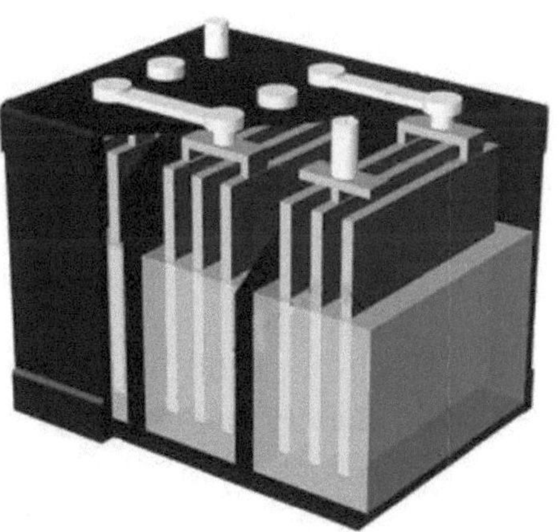

The battery is so called because it is an assembly of a number of components - each being an electrical cell. It can also be referred to as an accumulator, because it stores or accumulates energy. A 6-volt (6V) battery consists of 3 cells in series, each nominally of 2 volts. A 12-volt battery has 6 cells in series.

In each cell of the original Nimbus 6V battery there are a number of plates of lead and lead oxide. They are connected to the positive or to the negative posts. The cells are filled with an electrolyte, which is a 37% sulphuric acid concentrate.

The resting voltage of a fully charged battery should be 6.15 volts. This can be measured by a voltmeter connected across the two posts. If battery voltage is as low as 4 volts, it is of little use to charge it, as it is unlikely to recover a resting voltage of more than 5.5 volts as one of the battery cells must have failed. Riding with a battery in this condition will overload and damage the dynamo.
The battery needs to be disposed of.

Note that batteries are to be treated as environmentally dangerous waste.

The chemical activity that takes place in the battery (electrolysis) results in a reversible change in density of the electrolyte. The density rises under charging and falls under discharge.

The charging voltage must be higher (7.2 volts) than the resting voltage (6.15 volts) and when the motor is running it is the regulator which sees to that. See Section 3.3 (Voltage regulator)

Battery maintenance
The electrolyte level in a "wet" battery - that is one with removable caps on each cell - needs to be checked regularly, especially in the riding season.

Never use a lighter or a naked flame to check the acid level, as the gas given off by the battery is a mixture of oxygen and hydrogen - a highly explosive mixture.

In every cell, the electrolyte level should be high enough to just cover the plates. If it is not, it is because some of the water has evaporated and needs to be replaced. Always top up with water, never with acid. Use distilled or demineralised water, and do not overfill a cell.

In warmer times of the year, a little evaporation is to be expected, but if frequent topping up is needed, this could be a sign of over-charging, meaning that the regulator may need adjustment. But first, check that earth connections are sound, both the battery negative post to the frame and the regulator earth connection to the frame.

If a battery needs charging out of the motorcycle, a 6V home charger will normally bring the voltage up to the required level. But take care not to leave the charger connected for too long. Over-charging may cause overheating and gassing - giving off the explosive mixture of gases previously mentioned, and concentrating the electrolyte, possibly starting a process of decay in which lead falls from the plates to the bottom of the cell. This is often what causes a cell to "turn off", rendering the battery useless.

It is possible, at moderate cost, to buy a 6V "smart charger" which will not over-charge the battery as it will switch itself off when charging voltage reaches the 7.2 volts maximum.

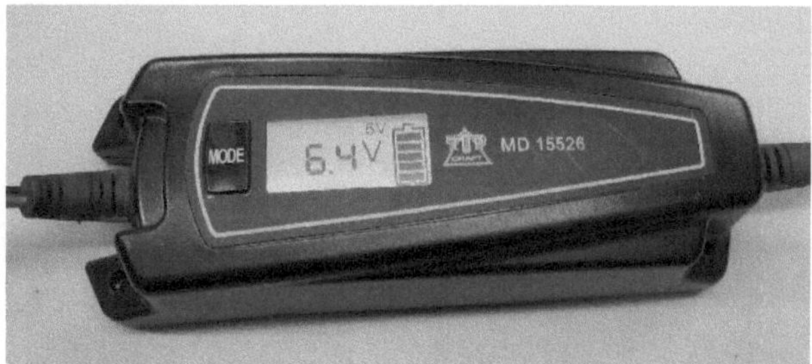

There are various types of low maintenance and maintenance-free batteries. Some are sealed but are otherwise much the same as the traditional wet battery. Take care to follow installation instructions, as some will quickly fail if not installed vertically and on their base.

There are also new technology batteries using lead alloy plates which extend battery life.

Today, batteries are used for many, some new, purposes, and the technology in this area is rapidly developing in ways which may find application in motorcycles.

3.2 Dynamo

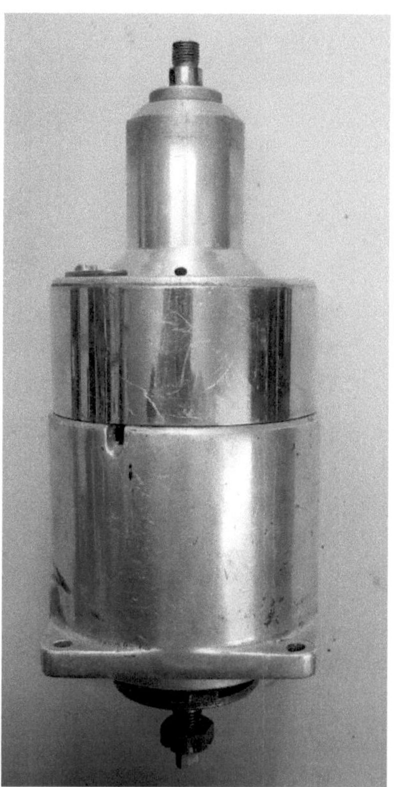

In 1934 - and in that year only - the Nimbus C was equipped with a 3-brush dynamo, entirely different to the dynamo fitted in 1935 and in all subsequent years.

The 1934 dynamo was not without its problems, and the manufacturer issued the following advice to owners:

Electrical power is generated by a vertical dynamo in front of the cylinder block. The dynamo is built on the well-known simple and robust 3-Brush system. It must not be operated without a battery as in that case the voltage will rise such that lights and the ignition coil will be damaged. Therefore always ensure that the battery and its connection to the dynamo is in good condition. The third brush can be adjusted by sliding the semicircular plate on the dynamo. Marks 1 and 4 indicate the charging current in amperes. Normally the brush is set to 4 amps, but for long runs without the use of horn and lights the ampere setting can be set lower. The dynamo is connected through a relay to the frame (negative terminal of the battery). The relay is closed by engine oil pressure.

If the dynamo is not charging, the cause may be in the relay, failing oil pressure, dirty commutator, worn brushes, insufficient spring pressure on brushes, or loose connections. The relay function can be checked by connecting the relay cable directly with to earth. If the dynamo then delivers a charging current, then the relay is at fault. The contacts may need reconditioning.
The commutator should also be carefully cleaned with fine emery cloth.

Owners, dealers, and the factory experienced so many problems relating to this version of the dynamo that many were replaced by the new version introduced in 1935. For this reason, there are few of these early dynamos to be found today.

Dynamo 1935 and later

The dynamo fitted to the Nimbus-C from 1935 onward is a DC (direct current) generator.
It has a wire-wrapped armature rotated by the motor. The armature wiring is arranged in a number of separate armature coils.

Each armature coil is connected to a pair of brass segments comprising the commutator and thence to two carbon contacts - the dynamo brushes. One dynamo brush is connected to the "D" terminal on the dynamo right side, and the other is earthed through a bare copper wire to one of the brush holder fixing screws on the other (left) side.

The armature rotates in the magnetic field of two electro-magnetic pole shoes attached within the dynamo casing. Around the two pole shoes are the field windings - the field coils.
Without the armature rotating there is only a relatively weak magnetic field between the two poles - the so-called residual field.

When the armature is rotated, a voltage is generated in the armature coils. (As in H. C. Ørsted's description of electromagnetism). This voltage depends on the strength of the magnetic field and the speed of rotation of the dynamo.

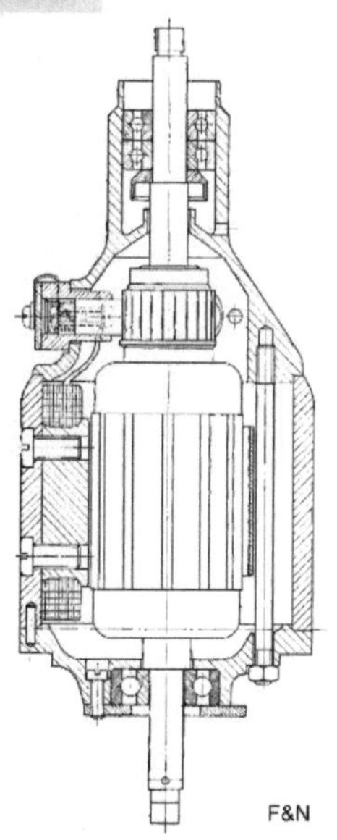

This is how the dynamo, battery, and regulator interact:

Right

Left

When the key on the combination switch is turned to its second position in order to start the motor, it is not just the ignition coil which is connected to the battery 6-volt supply. The regulator is also connected, and through a set of contacts in the regulator, so are the dynamo field coils via regulator terminal "F". See Section 3.3 (Voltage regulator).

Once the field coils are connected to the 6-volt supply in this way, the magnetic field is greatly increased, and the voltage produced in the rotating armature coils - the dynamo output - is correspondingly increased, increasing further with rising engine speed.

This is a process that is self-reinforcing, and at 4500 rpm, where the motor develops its maximum power, the voltage could rise to 24 to 26 volts.
However, this doesn't happen as the regulator limits the current to the field windings, limiting the strength of the magnetic field and thereby controlling the dynamo output voltage to a maximum of 7.2 volts.
That is what is needed to hold the battery charged and to enable the lighting system to perform at its best. (If a 12V electrical system is operating, the equivalent maximum is 14.4 volts).

Dynamo maintenance (1935 and later)

The only maintenance the dynamo requires relates to the electrical connections of the brushgear and the condition of the brush holders and the brushes themselves.

On the left side of the dynamo access to the brush holder is improved if the camshaft oil feed and drain pipe is first removed. The brush holder can then be withdrawn and the free movement of the brush in its holder can be checked. All oil or carbon dust contamination should be cleaned away. If any metallic particles are present this indicates armature damage and the dynamo should be removed for inspection and possible repair.

Before refitting the left side brush holder clean the commutator using a soft cloth lightly soaked with kerosene. Use the kickstarter to rotate the commutator as needed.

When refitting the left side brush holder note that there must be an earth connection from the brush via a short bare copper wire to one of the brush holder fixing screws.

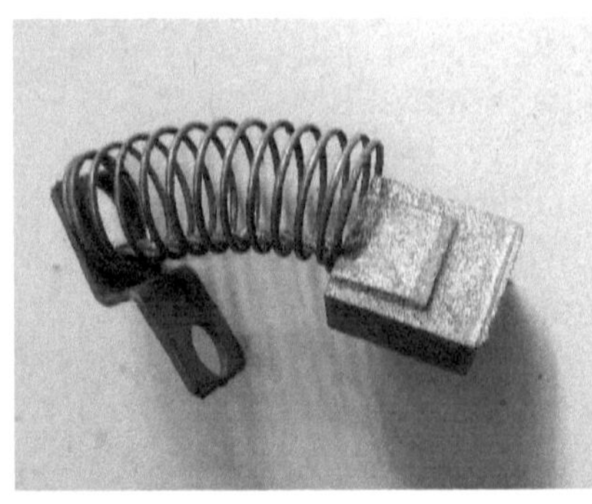

On the right side *leave the brush holder in place* and remove ONLY the cover of the brush holder. The brush holder must NOT be withdrawn as it is internally connected to the field wiring.

Mark the connecting leads "D" and "T" and take out the brush "D" together with its lead.

Check the brush to see if it has developed any wear which might prevent the spring from moving the brush into good contact with the commutator. If this is the case, replace it or both brushes.

Note that if this right-side brush holder with its connection to the field coils has been withdrawn, there is a risk that the connecting lead may have been displaced or the insulation damaged, critically affecting dynamo function. The dynamo may need to be removed, dismantled and correctly reassembled.

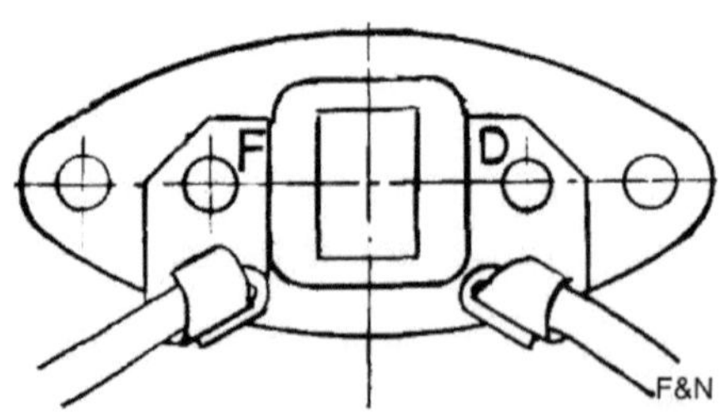

Dynamo testing and repair

If the dynamo is to be tested, it can be done with the dynamo (1) installed - that is, fitted to the motor, or (2) uninstalled, or (3) uninstalled and dismantled.

(1) Testing an installed dynamo

This is testing with the dynamo in position on the motor. Needed for these tests are a *voltmeter*, an *ammeter*, and two *'jump leads'* (short lengths of cable with crocodile clips on both ends).

Note that testing in this way results in maximum load on the dynamo. Take care to avoid testing for any longer than is necessary otherwise the dynamo may be damaged.

First, label the "D" and "F" cables, so they can be correctly reconnected, then remove both leads from the brush holder on the right side of the dynamo, but leave the "D" brush in position.

Measuring residual voltage

Residual voltage is the voltage generated in the armature when there is no electrical supply to the field coils and therefore only a residual magnetic field.

Connect the voltmeter red lead (positive) to the "D" brush holder terminal, and the voltmeter black lead (negative) to a good earth point on the frame, or to the battery negative terminal.

Start and *idle* the motor until the voltage reading stabilises. This should be at about 1.5 volts. If there is no reading, or if the voltmeter shows below zero, the dynamo needs to be polarised. The process is described below.

(See ***Polarising the dynamo.***)

After polarising, test again. If there is still not a reading of about 1.5v there is a fault in the armature and specialist attention will be needed.

Measuring maximum voltage

As above, connect the voltmeter red lead (positive) to the "D" brush holder terminal, and the voltmeter black lead (negative) to a good earth point on the frame, or to the battery negative terminal.

Use a jump lead to connect the brush holder "F" terminal to an earth point. Start and run the motor, gradually increasing speed until the voltage reading stabilises. This should be at 24 to 26 volts. If this is not the case, there is a fault either in the armature windings or in the field coils.

To determine where the fault lies, the dynamo must be dismantled.

See ***Testing an uninstalled dynamo.***

Measuring charging current (amperage)

Using two separate jump leads, connect dynamo "D" terminal to the battery positive post, and the dynamo "F" terminal to the battery negative post or a good earth point. Connect the ammeter to "D" and to the battery positive post.

Start and run the motor and check the ammeter. It should read about 10 Amps - if it does not, this indicates a fault in the armature windings or in the field windings.

To determine which is the case, the dynamo must be removed and dismantled.

See ***Testing a dismantled dynamo.***

Polarising the dynamo

Dynamo polarity can be lost following battery or regulator replacement or other electrical work. If the battery, regulator, and wiring harness appear to be in order but the charge light does not go out as the engine speed increases, it could be the case that the dynamo polarity has been has lost (or reversed), and repolarising is necessary.

The process is as follows:

> Run the motor, and with a short jump lead connect - just for a few seconds - regulator terminals "D" and "B". The charge light should go off and remain off when the short jump lead is removed.

If the charge light does not stay off after this procedure, the problem is not dynamo polarity, but a faulty armature.

(2) Testing an assembled dynamo which is not installed

This testing procedure is normally done before a refitting a dynamo which has been removed for repair. A 6 volt battery and 3 short jump leads with crocodile clips on both ends will be needed.

Connect the battery negative post to the dynamo casing. Connect the battery positive post to terminal "F" of the brush holder. If the dynamo, perhaps with a little help, rotates slowly *against* the direction indicated by the arrow stamped on the edge of the casing, then the field pole windings are in order.

Connect the battery negative to the dynamo casing and to brush holder terminal "F".

Connect the battery positive to brush holder terminal "D". If the dynamo, again with a little help, rotates rapidly in the *correct* direction, as shown by the arrow stamped on the edge of the casing, then the dynamo is (normally) in order. and ready to fit.

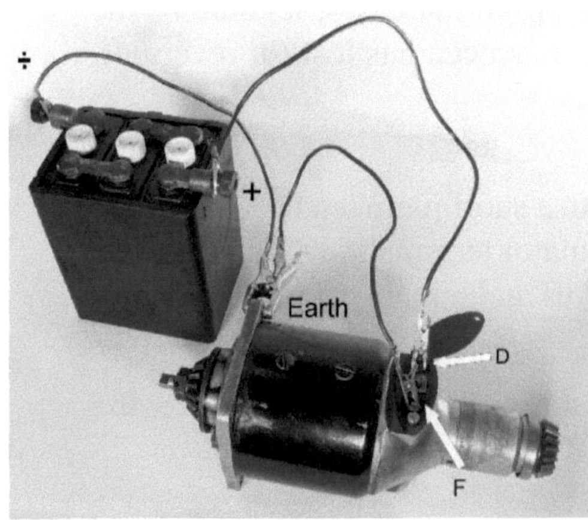

Testing the output voltage requires that the dynamo be fitted to the motorcycle or run on automotive electrical workshop testing equipment.

(3) Testing an uninstalled and dismantled dynamo

More of the dynamo components can be tested when it is dismantled.

Checking for continuity in the field windings

The field windings are mounted around the two magnet pole shoes which are screwed securely to the cylindrical dynamo body.

The field windings are fitted with a connecting insulated wire and two leads with terminals marked "D" and "F".

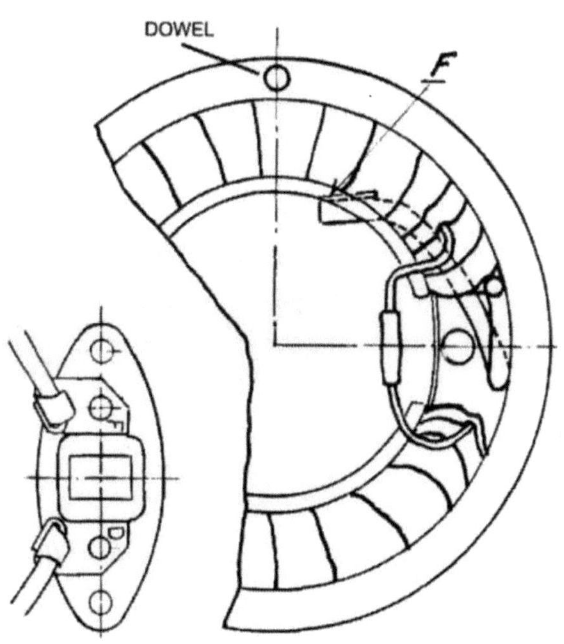

With a battery and test light, check for continuity between these two terminals.

There should also be continuity from one of these terminals and the dynamo casing.

If there is no continuity, check for any visible wiring or connector damage and correct it.

F&N

Testing the magnetic field

Hold for example, a screwdriver loosely between the two field poles while applying voltage across the field winding leads. The field is in order if the screwdriver is attracted by both magnet shoes.

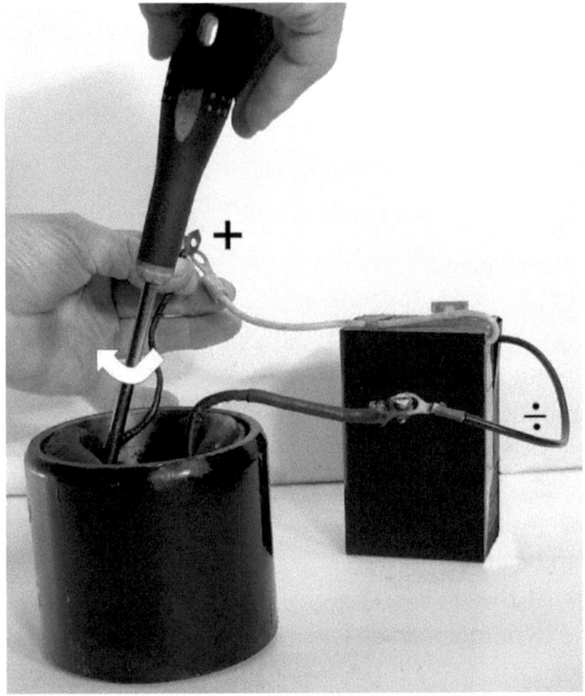

Checking for continuity in the armature windings

Again, use a battery and a test light.

Connect a short jump lead to the battery negative, and the test light to the battery positive.

Test whether there is continuity between every separate segment of the commutator and one - only one - of the others. (It will be the opposite one).
If this is not so, the armature needs repair.

Be aware that this test cannot detect all possible faults in the armature winding. Using specialist equipment (an "armature growler") an automotive electrician can do further testing.

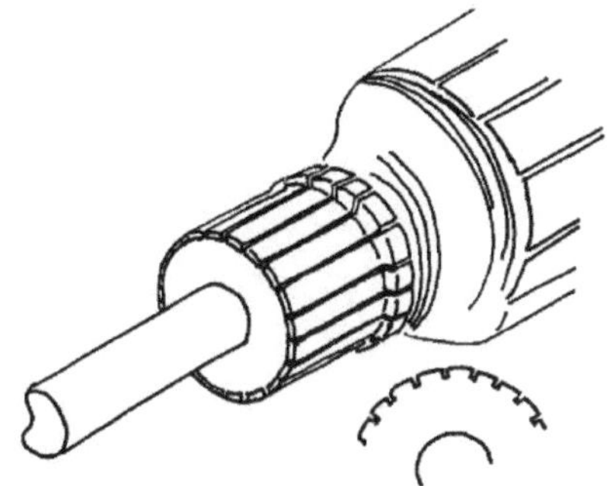

3.3 Voltage regulator

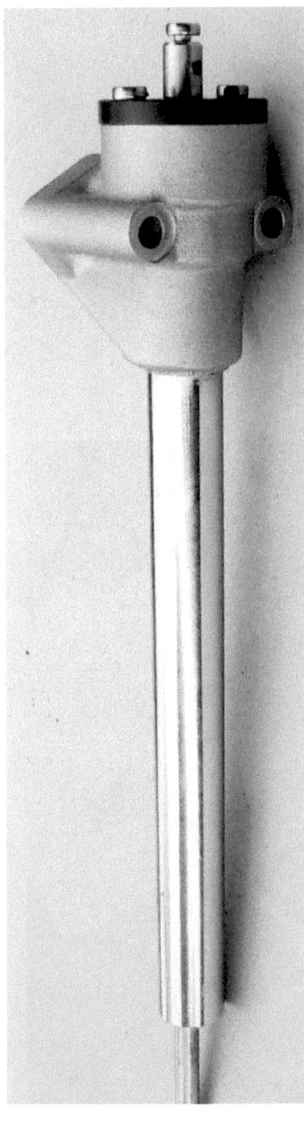

The full name of this device describes its function.

The Nimbus-C regulator adjusts the dynamo output voltage taking account of battery condition and electrical system demand.

The 1934 Nimbus-C had only a single oil pressure operated relay which served as a cut-out, disconnecting the battery from the dynamo when the motor was not running. Dynamo output was not automatically regulated and had to be set manually, to suit riding conditions. This system was used only in 1934 - 35.

Since then, electro-mechanical voltage regulators have been fitted to all Nimbus-C machines, initially devices of the factory's own design, later supplied by Bosch.

In recent times, electronic regulators have been developed. This type is entirely maintenance free, and any malfunction will require electronic repair likely to be out of scope for everyday owners.

Beneath the cover there are two relays: one is *the cut-out*, and the other is *the voltage regulator*. They have different functions but are similar in appearance and in construction, each using an electromagnet to attract an armature, which turns a switch on or off. The cut-out switch is off when the ignition is off, disconnecting the battery from the dynamo.

1935 – 1947

1948 - 1954

When the ignition key is first turned on, the charge light is activated because it is permanently earthed via a connection to the dynamo field winding and is now connected to the 6.15 volt battery supply.

Bosch
1954 - 1959

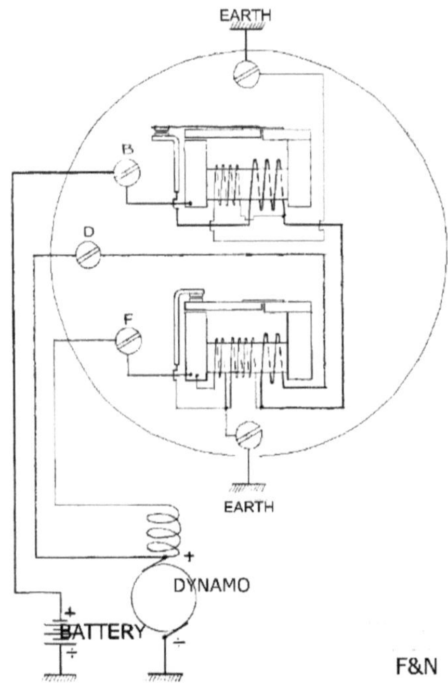

Once the motor is running, the dynamo armature is rapidly rotated in the magnetic field generated by the now powered field coils. An increasing voltage is developed in the armature windings. When this reaches approximately 7 volts, the cut-out relay switches on the connection to the battery so that it receives charge, and the charge light goes out because the charging voltage is higher than the battery voltage.

To prevent over-charging, the voltage regulating relay opens at a set voltage disconnecting charging current to the battery. When the battery voltage falls below the set level the regulator contacts close, switching charging current to the battery again. This process is continually repeated, keeping the battery correctly charged.

As long as the contact points are kept clean and dry the relay will work as it should.

Regulator maintenance:

It is just a matter of keeping everything under the regulator cover as clean and dry as possible. If the motorcycle has been exposed to damp for some time, it will be a good idea to check the contacts removing any contamination using very fine abrasive paper, being sure to blow all grit well away afterwards.

If it is suspected that the charging voltage is set too high, causing the battery to frequently need topping up, the voltage regulating relay can be adjusted and the charging voltage lowered.

This requires some electronic equipment and may not be a task for the owner.

Section 4: Ignition system

4.1 Ignition Coil

The ignition coil and the distributor are combined in one unit. The ignition coil consists of two separate and concentric windings around a soft iron core, sealed in a capped bakelite casing which has four brass rotor contacts riveted to the base. Each is connected to an external socket taking a high tension lead (HT lead) to a spark plug.

Of the two ignition coil windings the primary coil is connected to the battery through the ignition switch and the voltage in this coil is that of the battery - a little more than 6 volts. Current to the primary coil is switched on and off by the contact breaker points, inducing a current in the secondary coil.

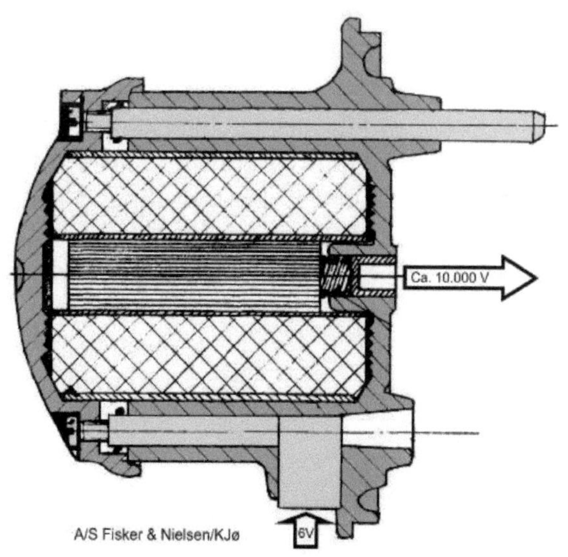

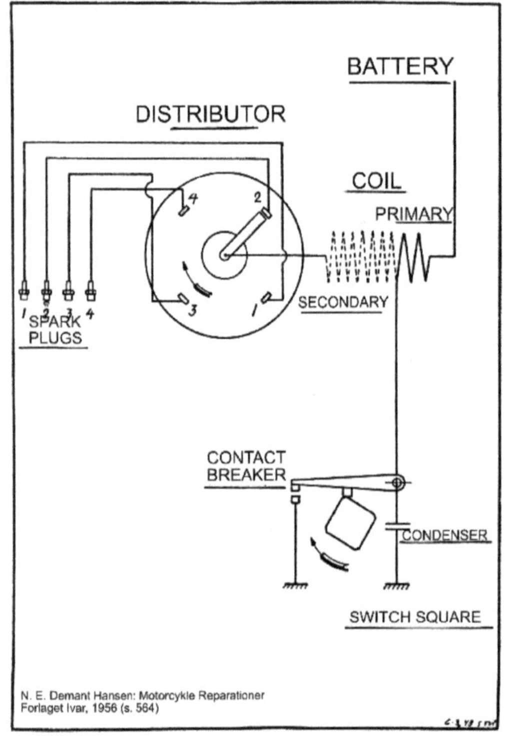

Because this coil has many more turns than the primary the induced voltage is very much higher - of the order of 10,000 volts. This high-tension current enables spark plugs to fire, maintaining the combustion process in the cylinders.

The primary coil is connected to a 6-volt supply in the distributor assembly by means of a brass probe embedded in the side of the ignition coil housing.

The secondary coil is connected to a spring loaded carbon brush which conducts the high- tension current through the distributor rotor to the four rotor contacts and so to the HT leads and the spark plugs. (See the diagram.)

Both the bakelite of the ignition coil housing and the insulated copper wiring of the coils may be 60 or even 80 years old. This means that the bakelite is probably brittle, with fine cracks that allow rainwater and dirt to penetrate. Wiring insulation may break down leading to short circuits within a coil, effectively reducing the number of turns. Moreover, there is evidence that copper wiring degrades over time and has a limited service life. Certainly in recent years it has been the case that there have been ignition coil problems and failures.

Typically what happens is that a Nimbus will start and run normally for some time, but when the ignition coil warms up to its operating temperature it begins to misfire.

The problem gradually gets worse and eventually the motor cuts out completely. After cooling down for an hour or so it can be started again, but the experience is repeated, just as before.

This is because the conductivity of copper decreases (its resistance increases) as the temperature is increased. So a defective coil which produces only a weak spark when cold, causing starting difficulty and rough running, will function less and less well as it heats up in use.

If these indications of a progressively failing coil are present, it is wise to have this confirmed by a workshop which has the equipment needed to fully test the coil, as poor starting and erratic running can also result from problems elsewhere in the system.

Ignition coil maintenance

Owner maintenance is limited to keeping the coil as clean and dry as possible, as should the coil be suspect, owner repair is not advisable.

First, the old bakelite housing is so brittle that it is likely to crack during attempts to disassemble it. Second, the coil was originally sealed in the housing with liquid pitch, now set quite hard, and very difficult to remove. A replacement or repaired coil has to be sealed in a similar way.

4.2 Distributor

The distributor base is a steel pressing which has a number of large and small parts fixed to it. Every part is essential to the correct working of the ignition system, so all rivets and other fasteners need to be secure, and where required, need to provide good electrical contact.

F&N

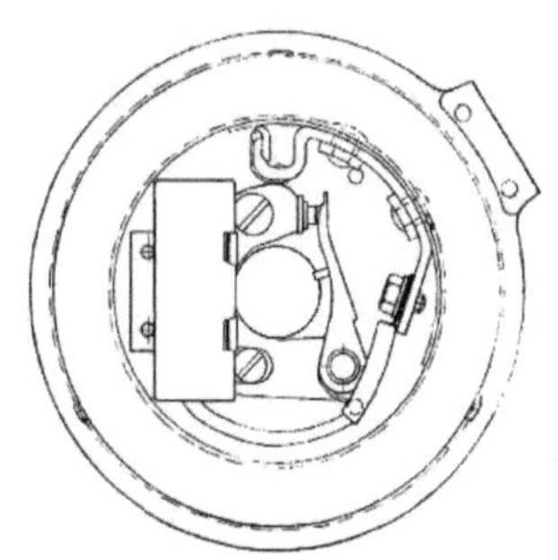

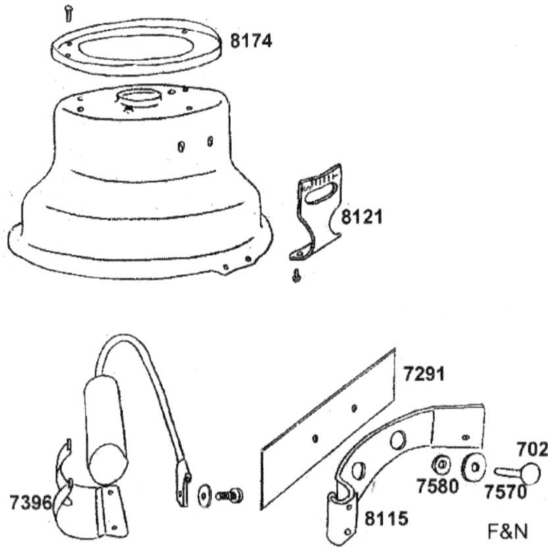

8174 Oil shield and 7396 Holder for condenser

Rivets securing both parts need to be tight, and the holder should have enough spring pressure to grip the condenser firmly.

8121 Timing indicator plate

This should also be riveted tightly, as it provides an earth connection by means of the external timing securing screw of the camshaft housing.

All other parts must be well insulated from the distributor base. Note that the ignition coil probe terminal bracket (8115) is isolated from the distributor body by the use of insulating washers (7570) and insulating bushes (7580) around the copper rivets (7027).

Distributor maintenance

A small amount of oil is always drawn past the oil shield into the distributor from the camshaft housing, and moisture can leak in from outside. It is occasionally necessary to attend to this.

This is best done by removing the coil and distributor from the camshaft housing and taking out the components which can be removed - the condenser and contact breaker set.

Slight moisture or corrosion around the copper rivets and insulating washers and bushes can result in electrical leakage - a partial short circuit in the distributor.
Such a condition cannot normally be detected by an everyday ohmmeter.

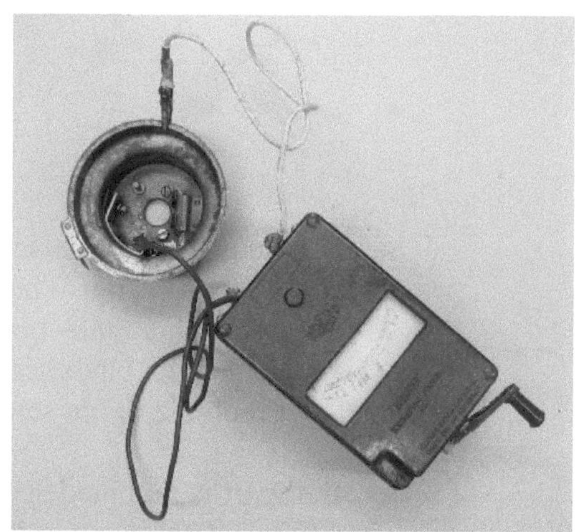

To be absolutely sure that the resistance between the terminal racket (8115) and the distributor base is infinite so that there is no shorting, a special-purpose insulation tester is needed (a mega-ohm meter).
Best therefore to have an electrical workshop check the distributor.

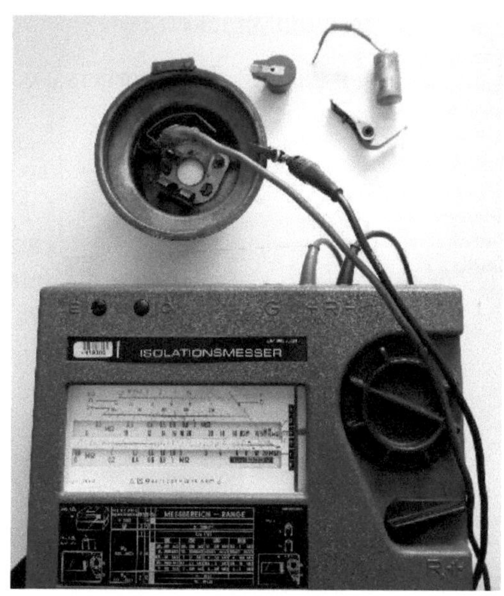

If there is a problem with the insulation of the terminal bracket the rivets can be drilled out, new insulating washers and bushes fitted, and the bracket re-attached to the distributor base using new fasteners - 4mm pop rivets or 4mm screws with locknuts.
At the same time all other rivets should be checked, and any loose ones tightened with a rivet-setter or similar tool.

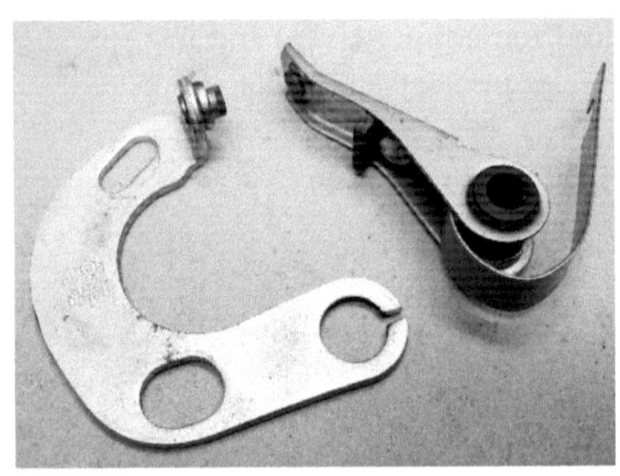

4.3 Contact breaker and Condenser

Contact breaker

The contact breaker, often referred to simply as "the points" consists of two separate parts.

One part - the "fixed" contact - is screwed to the base of the contact breaker housing. When this fixing screw is slackened, it can be moved slightly by turning the nearby eccentric adjusting screw with a flat blade screwdriver. This adjusts the points gap after which the fixing screw needs to be tightened.

The other part is mounted such that it can pivot about a fixed post (7397) and has a leaf-spring attached to the ignition probe terminal bracket.

On each of these parts is a contact point.

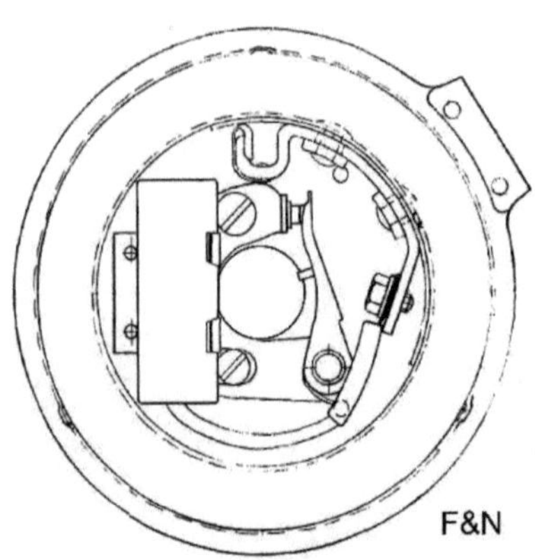

To adjust the points gap the distributor must be firmly in its place on the camshaft housing, with the camshaft rotated to a position in which the fibre heel of the moving point is resting on the one of the four "peaks" of the ignition cam.

Loosen slightly, and do not remove, the fixing screw of the adjustable point plate. Next, use the screwdriver to turn the eccentric adjusting screw such that the points gap is 0.7mm. The fixing screw should then be tightened.

Check that the points gap is also correct at each of the three other ignition cam peaks.

If it is not, it may be that the distributor is not pressed fully home into the camshaft housing, or the fixing screw is not tight.

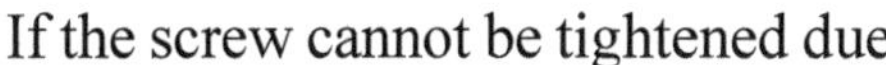

If the screw cannot be tightened due to a worn thread in the distributor base, this needs attending to - for example, a 4mm nut can be soldered to the back of the distributor base.

Contactbreaker maintenance

A contact breaker set has a limited life, but with a little care, it can be a long life. At a crankshaft speed of 4500rpm the camshaft is spinning at half that speed. And in every single revolution of the camshaft the contacts open and close 4 times. That is, 10,000 times in every minute. Wear is inevitable, and therefore so is contact breaker points replacement.

When fitting a replacement set, clean both contacts with a solvent to remove any protective coating, and perhaps lightly polish them with very fine emery paper.

Clean the tip of the rotor in a similar way.

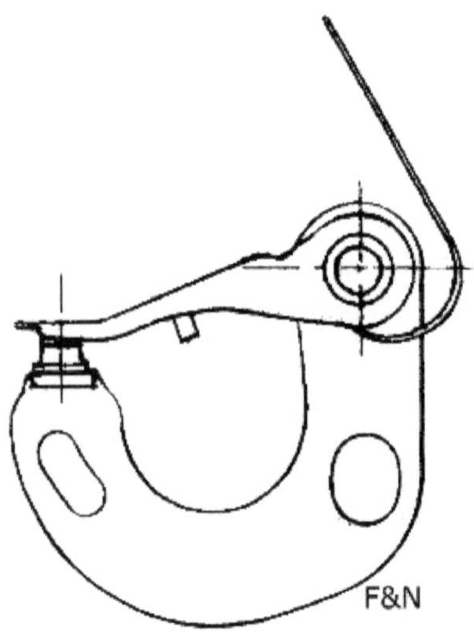

Special "contact files" are available, but are not recommended, neither for an initial cleaning nor maintenance. Some are too coarse for the fine work that is needed.

Arcing at the contact breaker points is not an uncommon problem, usually a result of riding with a defective condenser and/or a deteriorating ignition coil. Arcing means sparks between the points which drag material from one contact to the other. In time a peak will develop on one contact and a pit on the other. Misfiring and rough running will be the result.

Electronic ignition system

Recently, a totally electronic ignition system has been developed, which replaces the mechanical points assembly. It can be installed entirely within the distributor housing so that the conversion is not noticeable.
There has not yet been long-term testing of the reliability of this equipment.

Condenser

A condenser is also known as a capacitator. Its function is to receive a certain amount of electrical energy during the time when the contact points are closed, and to release this energy at the instant the points open.

Once the condenser is removed, it can be checked using a multimeter that can measure microfarads (μ), and for a condenser suitable for the Nimbus, this value should be between 0.26 and 0.34 microfarads. Discard the condenser If the reading falls outside this range.

A modern ceramic condenser is available and is suitable for use in the Nimbus-C. There are positive reports from long-term testing of this item.

Condenser maintenance

Condensers require no actual maintenance and are not repairable.
All that is needed is that the securing clip has a good metallic/electrical contact with the cylindrical case of the condenser. This should be checked occasionally and if it is not so, the clip should be bent a little so that is grips the condenser tightly.

4.4 High Tension leads, Spark Plug Caps and Spark Plugs

HT Leads

The spark plug leads, or high-tension (HT) leads, conduct a current of thousands of volts.
They have to be in good condition - clean and undamaged.

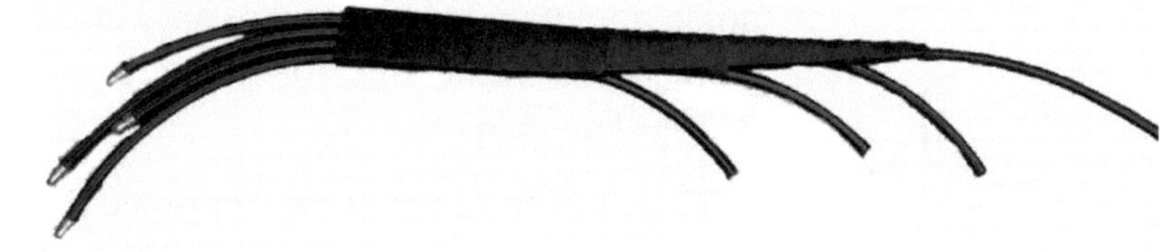

Early design

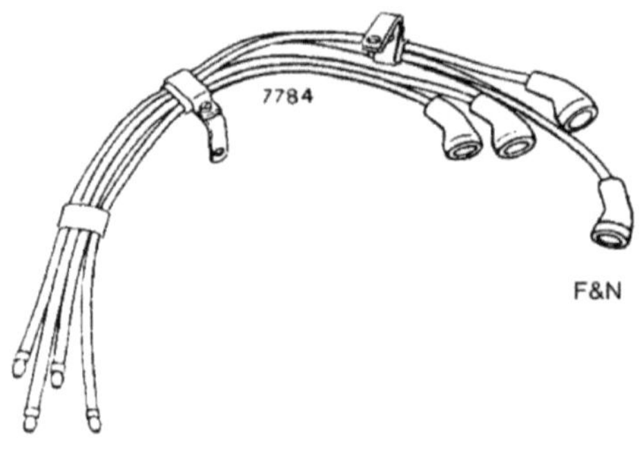

The wire core of an ignition cable consists of tightly twisted thin copper strands and they must all make good contact in the high tension output sockets of the distributor.
If the electrical contact is poor, a spark may still be produced, but there may be an unnecessary load L on the ignition coil.

Late design.

HT leads Maintenance

Remove the complete lead set and check that each cable has all copper strands soldered to the brass terminal that plugs into the ignition coil.

Make sure that for each lead the connection to the spark plug cap is in order.

The rubber insulation of all leads must be undamaged, with minimal stiffening and no cracking. Leads should be pushed firmly into the correct distributor socket. If necessary, use a small screwdriver to press them into place.

Spark plug caps

A great variety of spark plug caps is available.
At one time they were required to be radio noise attenuated (suppressor caps) and that made them vulnerable to moisture due to the built-in resistor.

This provision is not strictly enforced today. Therefore it is recommended that the original Nimbus spark plug caps be used.

They provide a very secure connection between the HT lead and the spark plug.

In these original spark plug caps, there are two versions of the metal fitting which secures the spark plug cap to the spark plug. The version which has a small transverse wire spring is the more secure.
Both versions are somewhat susceptible to moisture and contaminant ingress which can cause corrosion and malfunction.

Rubber cover (10945) from 1956 can be slipped over all original Nimbus plug caps. Enquire at a Nimbus dealer.

Spark Plug Cap Maintenance

As mentioned, moisture and contaminants can be a problem. Therefore it is advisable as part of general maintenance to unscrew the parts which secure the exposed wire core of the cable to the cap and clean them thoroughly.

Spark plugs

Four identical and well-maintained spark plugs of the correct type are essential for trouble free riding.

Many different manufacturers produce spark plugs suitable for the Nimbus-C. The first consideration is that the thread is the correct diameter and type: 14mm metric fine (M14 x 1.25mm) and the thread length (plug reach) is 1/2 inch or 12.7mm.

It is easy to find cheap new spark plugs at swap meets and so on which will fit a Nimbus. But plugs vary not only in dimensions but also in heat rating, from “hot” to “cold”, to take account of the requirements of different engines and riding.

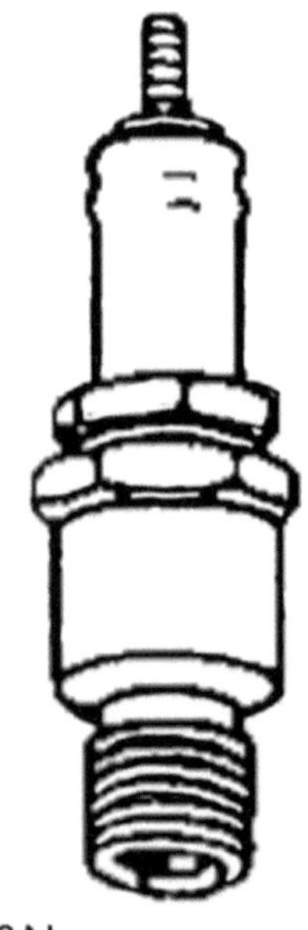

F&N

The heat rating is usually stamped on the plug body - but different makers code in different ways. The Japanese company NGK has specifically recommended their B4H spark plug for Nimbus motorcycles. The NGK coding system is: "B"= 14mm thread; "4" = heat rating, on a scale of 2 to 14; and "H"= 1/2" plug reach. Internet spark plug cross-reference sites list equivalent plugs from other manufacturers.

Information can be found on the Internet relating to the heat rating of spark plugs, and how the appearance of the plug tip can give useful indications of the general condition of the motor.

Nimbus dealers will supply spark plugs from various makers which they know from experience will be entirely satisfactory.

Spark plug maintenance

When removing or refitting the spark plugs care must be taken not to damage the insulator, which is porcelain, and brittle. Check carefully as a cracked insulator means the plug must be discarded.

Remove carbon or oily deposits from the tip of the spark plug with a fine wire brush, leaving the centre and side electrodes as clean as possible.

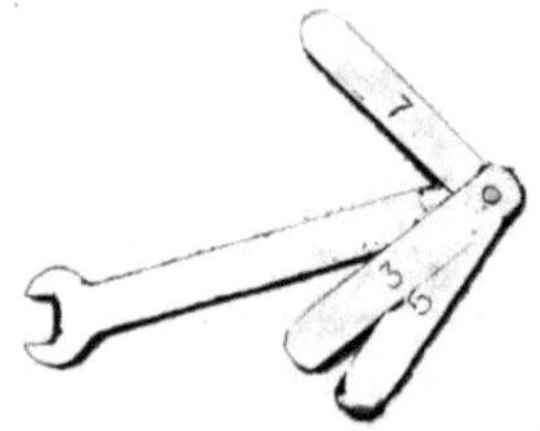

Set the electrode gap to the recommended 0.7mm (0.028") using a feeler gauge or plug gapping tool. Lightly grease the plug threads, and screw the plug in by hand, only using the plug spanner to apply final tightening.

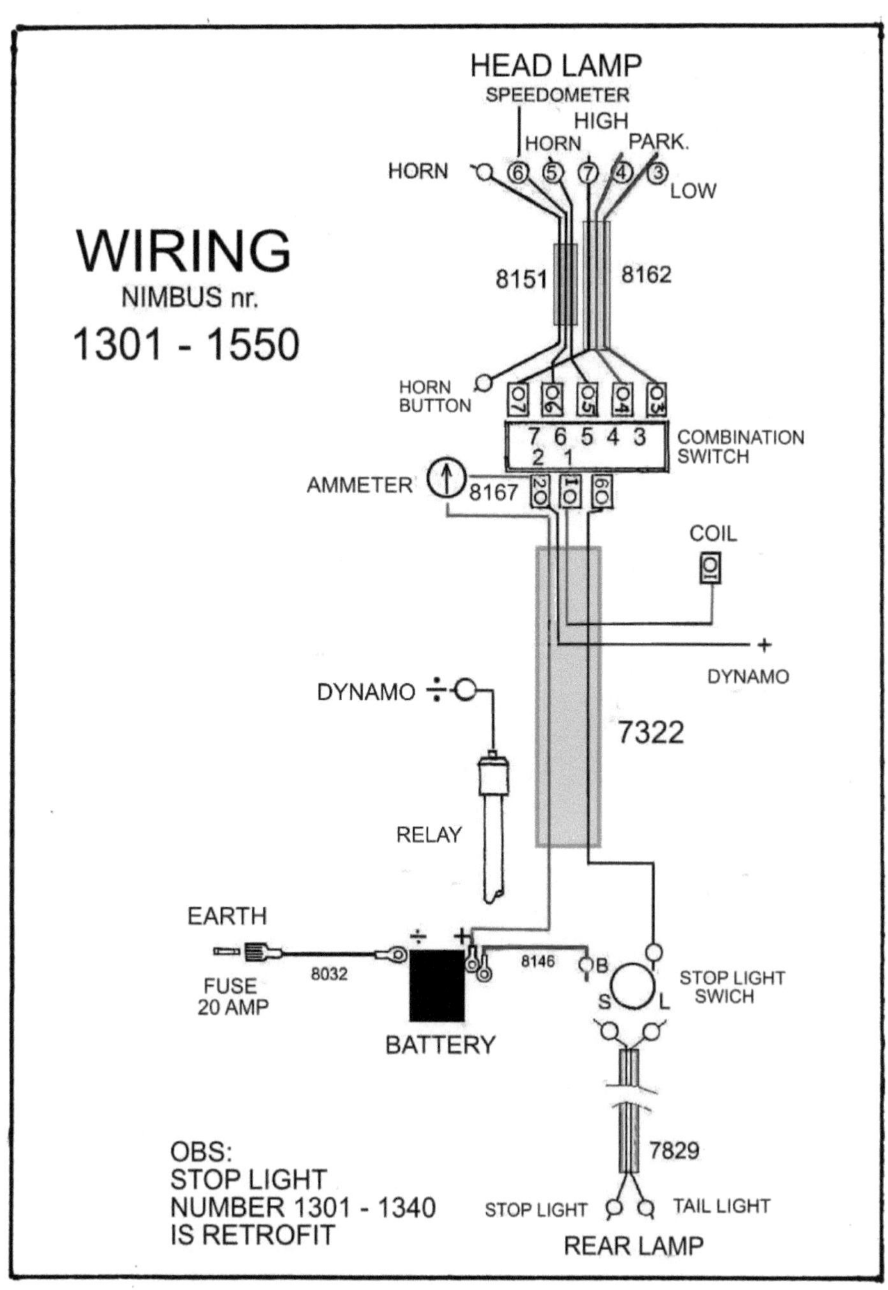
WIRING
NIMBUS nr.
1301 - 1550
HEAD LAMP
SPEEDOMETER
HIGH
HORN
PARK.
HORN
LOW
8151
8162
HORN
BUTTON
7 6 5 4 3
2 1
COMBINATION
SWITCH
AMMETER
8167
COIL
+
DYNAMO
DYNAMO
7322
RELAY
EARTH
8032
FUSE
20 AMP
8146
B
S
L
STOP LIGHT
SWICH
BATTERY
7829
OBS:
STOP LIGHT
NUMBER 1301 - 1340
IS RETROFIT
STOP LIGHT
TAIL LIGHT
REAR LAMP

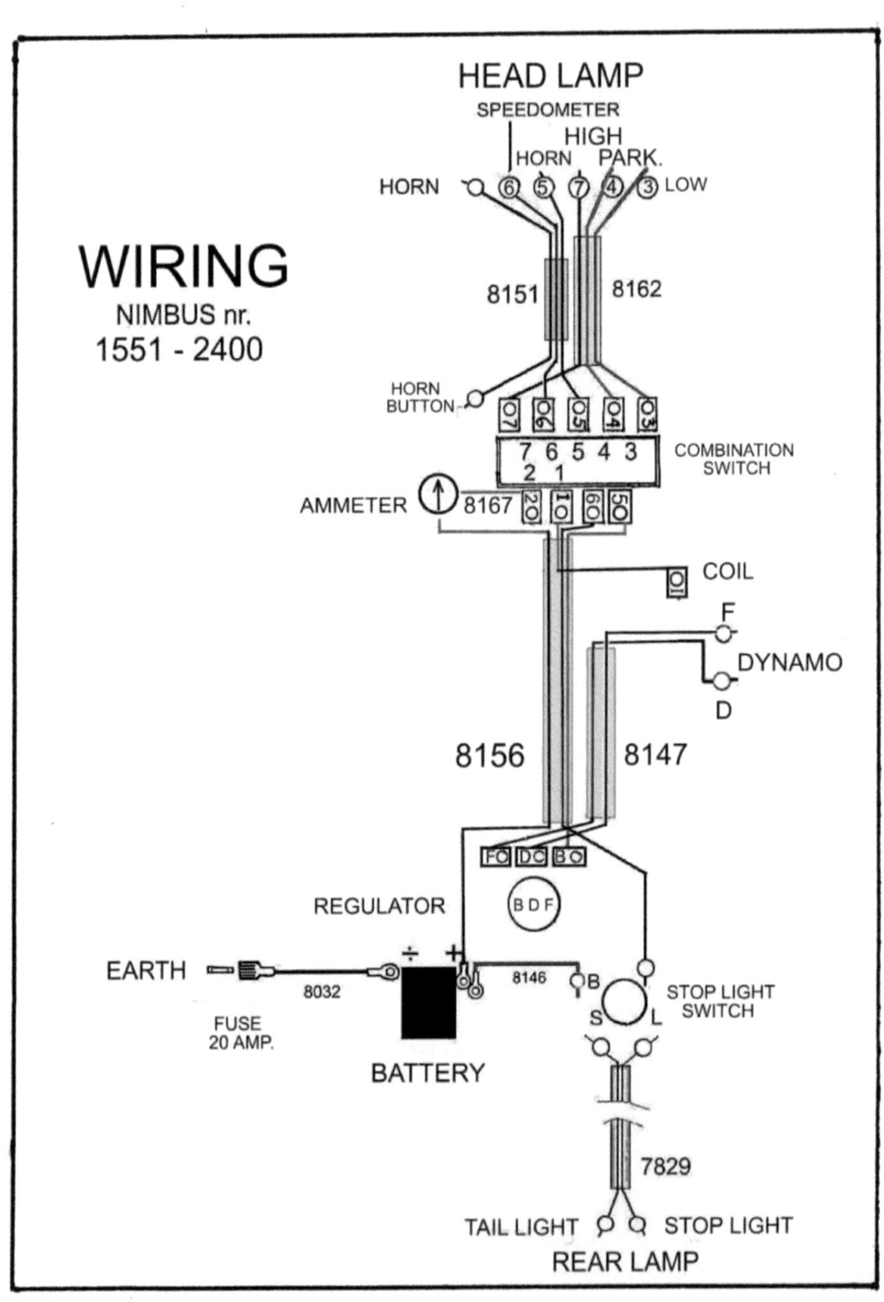
WIRING
NIMBUS nr.
1551 - 2400
HEAD LAMP
SPEEDOMETER
HIGH
HORN
PARK.
HORN
LOW
8151
8162
HORN
BUTTON
COMBINATION
SWITCH
AMMETER
8167
COIL
F
DYNAMO
D
8156
8147
REGULATOR
EARTH
8032
FUSE
20 AMP.
8146
B
STOP LIGHT
SWITCH
S
L
BATTERY
7829
TAIL LIGHT
STOP LIGHT
REAR LAMP

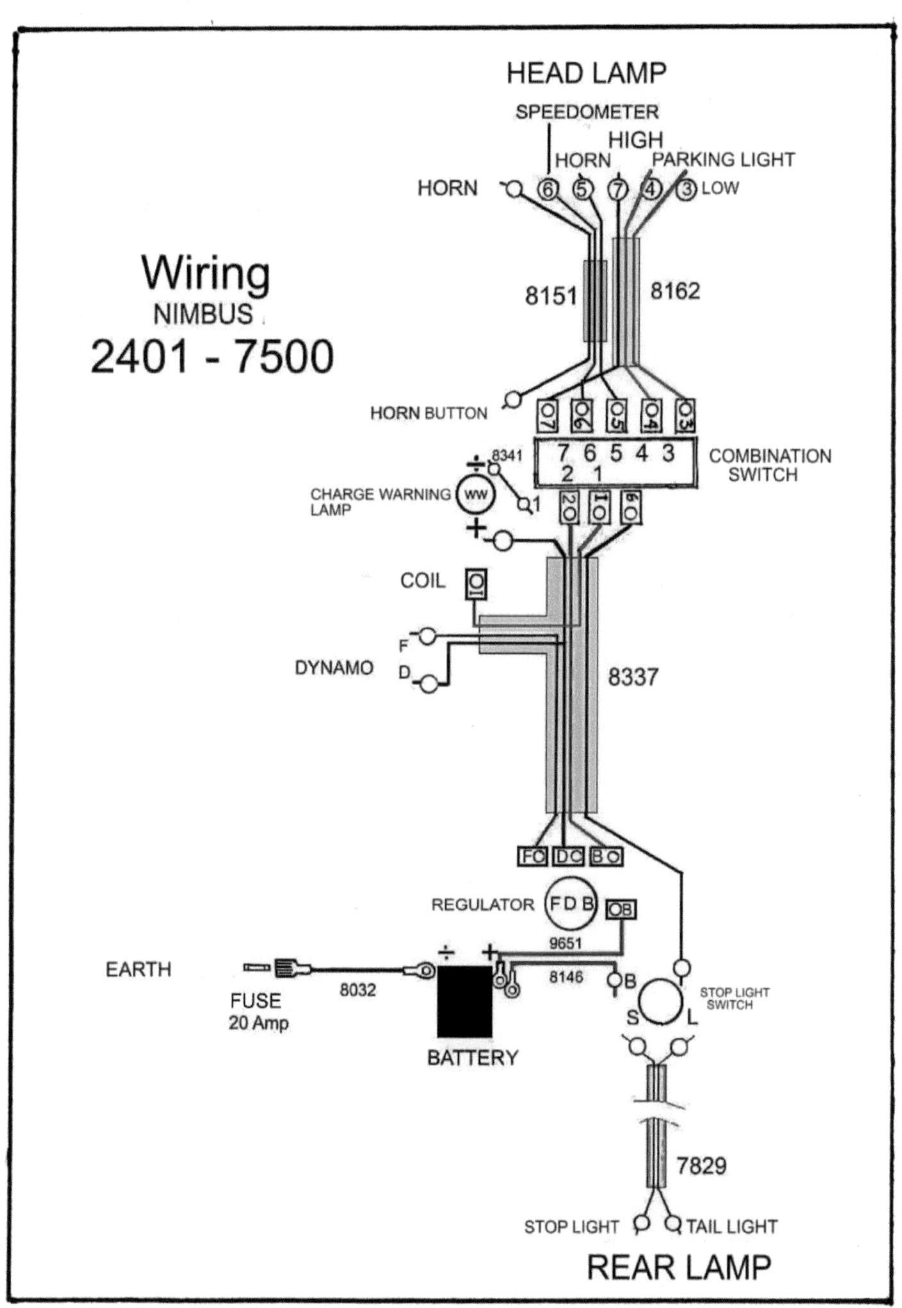
HEAD LAMP
SPEEDOMETER
HIGH
HORN
PARKING LIGHT
HORN
6
5
7
4
3
LOW
Wiring
NIMBUS
2401 - 7500
8151
8162
HORN BUTTON
7
6
5
4
3
7 6 5 4 3
2 1
COMBINATION SWITCH
8341
CHARGE WARNING LAMP
WW
1
2
1
6
COIL
F
DYNAMO
D
8337
F
D
B
REGULATOR
F D B
B
9651
EARTH
8032
8146
B
STOP LIGHT SWITCH
S
L
FUSE
20 Amp
BATTERY
7829
STOP LIGHT
TAIL LIGHT
REAR LAMP

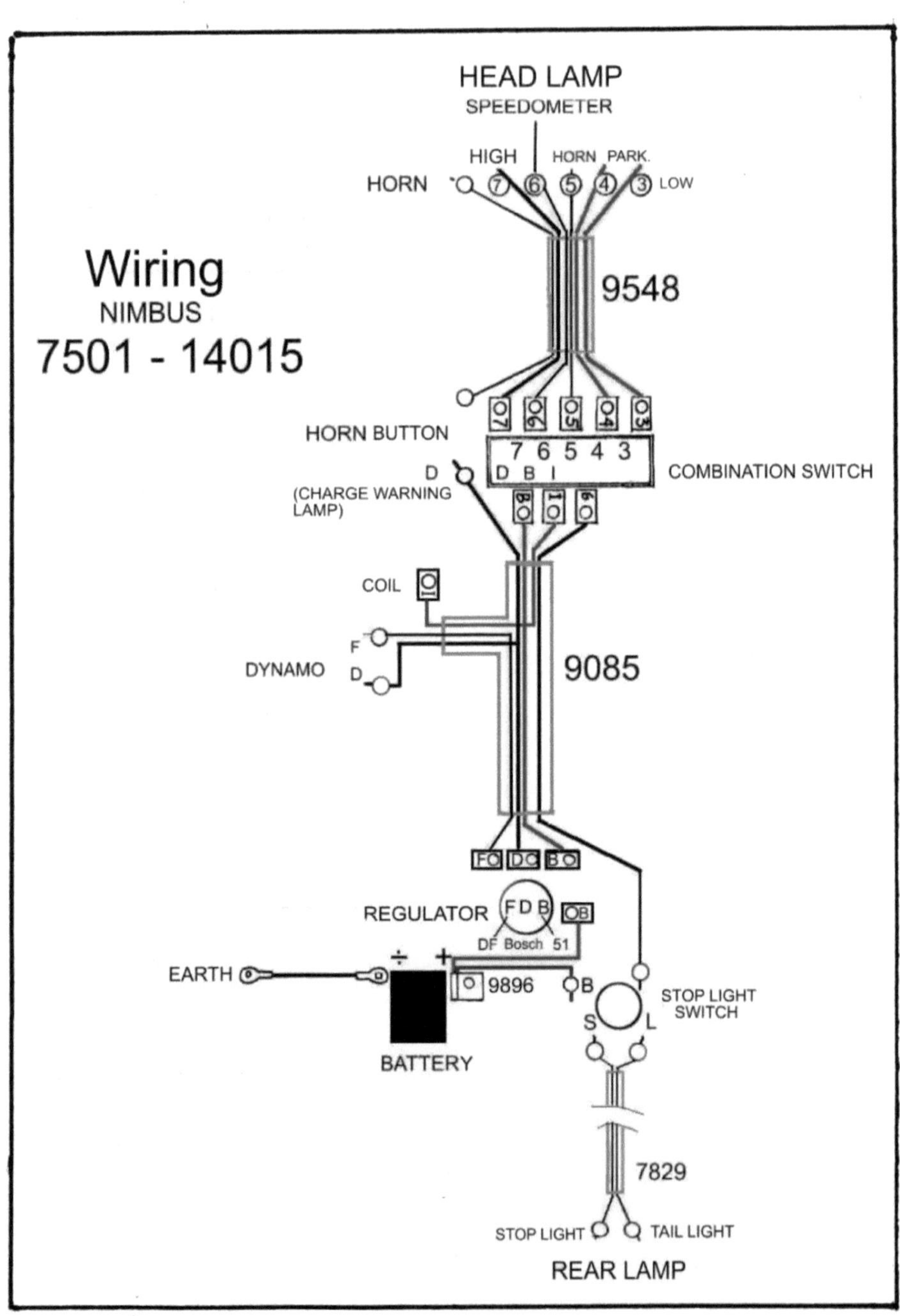
HEAD LAMP
SPEEDOMETER
HIGH
HORN
PARK.
HORN
LOW
Wiring
NIMBUS
7501 - 14015
9548
HORN BUTTON
7 6 5 4 3
D B I
COMBINATION SWITCH
D
(CHARGE WARNING
LAMP)
COIL
F
DYNAMO
D
9085
REGULATOR
F D B
DF Bosch 51
EARTH
9896
B
STOP LIGHT
SWITCH
S
L
BATTERY
7829
STOP LIGHT
TAIL LIGHT
REAR LAMP